MODIFYING YOUR FIBERGLASS BOAT

Jack Wiley

ISBN-13: 978-1514398074
ISBN-10: 1514398079

Printed by CreateSpace

CONTENTS

INTRODUCTION

This is a book of possibilities for modifying your fiberglass boat. By "modifying" I mean all or any of the following: reconditioning; refurbishing; restoring; equipping; giving a facelift to; improving such things as the appearance, comfort, and performance; and making any other changes to an existing fiberglass boat. By "your fiberglass boat" I mean one that you now own or one that you acquire in the future.

In 1975, International Marine Publishing Company published my book *Modifying Fiberglass Boats*, which was the fourth book I had written to be published. It was based on my building two wooden boats – an 11-foot pram while still in high school and a 30-foot trimaran ten years after that – living aboard a 21-1/2 foot sailboat for five years, and helping a number of people build their own world cruising sailboats from manufactured fiberglass hulls. For years I had been reading everything I could find about boats and boating and learning everything I could about boats and boating from other people who shared an interest in these subjects.

At that point in my life I thought I had learned everything about modifying fiberglass boats that I was going to learn. Now in 2015 at 78 years of age, I realize how wrong I was about this. For example, shortly after the publication of *Modifying Fiberglass Boats*, I purchased an old damaged and rundown 17-foot Dotline power cruiser and completely reconditioned it. This project was followed by five more years of living aboard boats and gaining a lot more experience modifying them.

What this means is that this book contains not only what I had learned about modifying fiberglass boats that went into the original book I wrote on the subject, but also everything I have learned since then. Even though the passion I once had for boating has now shifted mostly to RVing, I'm still using similar modifying techniques, only now they are being applied to RVs rather than boats. In other words, I'm still learning.

Why Modify a Fiberglass Boat?

Modifying a fiberglass boat, whether it is one you already own or one you have purchased especially for the purpose, is perhaps the best means of getting the most and best boat for the least money. New fiberglass boats are now in 2015 as I write this very expensive. However, between the time when fiberglass boats were first manufactured in the 1950s and now thousands of fiberglass boats were manufactured, and fiberglass hulls and other moldings have proven to have a long lifespan. The hulls on many fiberglass boats made in the 1950s and 1960s are still sound and useable. Old fiberglass boats from these and later years are readily available today, including rundown and damaged ones. By making a careful selection, you can purchase one of these that can be restored in a reasonable time at a total cost far below that of an equivalent boat in good condition. If you purchase a fiberglass boat that is already in good condition, you can modify it to make it even more suitable to your wishes and needs.

Here are some common reasons for modifying a fiberglass boat:

To make the boat safer. Just because a boat is manufactured and marketed does not necessarily mean that it is seaworthy. Many such boats have major safety deficiencies. By modifying, these shortcomings can often be remedied. In some cases, modifications can make a boat safe for a purpose more demanding than that for which the boat was original intended.

To make the boat more comfortable. Most stock boats seem to be designed for some average person or family that doesn't exist. There seems to be, for example, a compulsion to cram as many berths as possible into the boat, regardless of whether any of them is adequate. Six berths, all of them uncomfortable, are typical for a 25-footer. But there are many boat owners who would rather have two or three good berths, and more storage space. And what boat owner hasn't wanted a shelf here or there, or a rack?

To improve performance. This may mean increasing the speed of the boat or making it easier to handle.

To improve appearance. The concept is similar to that of customizing a car. However, for the boat it's extremely important that the modifications not affect the safety of the craft. Even with this restriction, though, many modifications are possible. Many owners want a boat that doesn't look like hundreds of others from the same mold.

Why Do It Yourself?

The most obvious answer is to save money. But it's not the only one. Two pitfalls are common when you have work done by a boatyard. First, it's easy to get really socked dollar wise. Second, having the job done the way you want it and when you want it is nearly impossible. Many, perhaps most, workers in boatyards can scarcely be referred to as "boat builders." To use an old cliché, they could care less. They only do enough to get by. On the other hand, many amateur builders exhibit better workmanship because they can afford the time to take pride in what they are doing.

This Book

This book is aimed at do-it-yourselfers. While you do not necessarily need any previous experience working on boats, you do need to be handy at working with tools. Although it is not necessary to have done any fiberglassing before (the basic techniques for doing this are covered in this book), I am assuming that you already possess basic skills in woodworking and metalworking. If you do not have these skills, this book is probably not for you.

The following chapters provide the information needed for modifying all types of stock fiberglass boats. A special feature of the book is the practice projects, which allow you to practice with scrap material when attempting new techniques. This is extremely important, as many of the techniques require skill. Too often, boat owners will try unfamiliar techniques on their boats, with unsatisfactory results. Also important, as emphasized throughout this book, is a thorough understanding of the job before attempting it on the boat. Quality workmanship demands this.

1

FIBERGLASS BOATS

Before proceeding to the fundamental techniques of modifying fiberglass boats, which are covered in the next chapter, let's take a close look at stock fiberglass boats in general, both from the owners' and manufacturers' points of view.

Fiberglass

As far as stock boats are concerned, fiberglass has taken over most of the market. The hulls, and in many cases other components, of the vast majority of boats manufactured, sold, and used today are molded fiberglass, and this has been the case for many years. There is a continuing trend for manufacturers of wood, plywood, aluminum, and steel boats to switch to fiberglass. Ferrocement, in spite of its publicity, accounts for only a small percentage of the stock manufactured boats.

What is commonly referred to as "fiberglass" is actually a fiberglass-reinforced plastic, which can be molded to any desired shape and which cures to a composite material of great strength and durability. Glass fibers are one of the strongest reinforcing materials known and are found most frequently in manufactured boats, though other materials can be used.

In most cases, the plastic used is polyester resin, which saturates and surrounds the glass fibers and cures to a permanent hardness. Although epoxy resin has superior strength and bonding qualities, it is more expensive and difficult to use, and therefore not commonly used in manufactured boats.

The laminate formed from fiberglass and polyester resin has high impact strength and resilience, and, pound for pound, is stronger than steel. Glass fibers and liquid resin can be formed in a mold to any desired hull form, making a one-piece shell with no seams or fasteners.

Construction

There are three basic methods used in manufacturing fiberglass hulls: hand lay-up, spray-up, and matched-metal die molding.

In the hand lay-up and spray-up, the hulls and components are built from the outside in, and the laminates are cured at atmospheric pressure. First, a pigmented gel coat is sprayed in the mold cavity, and then glass fibers are put in and saturated with resin. In the hand lay-up, layers of glass fabric and mat (glass reinforcing materials are discussed in the next chapter) are used. In the spray-up, chopped glass fibers and resin are sprayed into the mold with a "chopper gun."

In matched-metal die molding, a glass-fiber preform is made on a shaped vacuum screen, and is then placed over a male mold in a press for cure under heat and pressure. To date, this method has been limited to relatively small boats.

All three methods have been used to produce both high- and low-quality hulls and components. Important factors to consider, regardless of the method, are soundness of design, quality of materials, and care and skill in fabrication.

Advantages for Owners

We will first consider some of the reasons for the popularity of fiberglass over other boatbuilding materials, from the point of view of boat owners.

Durability. Fiberglass boats have proved to be durable. A quality hull will last many years. A number of the first stock pleasure fiberglass boats built in the forties still exist and their hulls are in good condition. Even in the most demanding marine environment, fiberglass won't rust or corrode. It's immune to electrolytic action. A high-quality, well-designed fiberglass boat is strong and damage-resistant, and should last almost indefinitely.

Value. Because of its durability, a high-quality fiberglass boat will give many years of service and will have a high resale value. Fiberglass hulls and decks require a minimum of scantlings, which results in more cabin and cockpit room than is available on a wood or metal boat of equal exterior dimensions. Maintenance generally costs much less than boats of wood or metal and entails less time in the yard, which means more time in the water.

Maintenance. Fiberglass boats do require maintenance, but most of it is cosmetic rather than for preventing deterioration. Marine growths have not been known to harm a fiberglass hull, but barnacles and other fouling organisms attach themselves to it, so antifouling paint is needed. However, neglect in this area will not cause serious damage, as is often the case with wood and metal boats.

Repair. Fiberglass boats are often advertised as being easy to repair, but this depends on many things. When fiberglass is fractured, the damage is general localized with edges fair and true. This makes repairing easy, but fairing and matching gel-coat finishes may be difficult. In most cases, assuming the damage is the same, repair is easier on a fiberglass boat than on a wooden one.

Selection. By far the largest selection of boats, new or used, is in fiberglass. Everything from prams to large offshore boats is readily available, including runabouts, houseboats, powerboats, and sailboats.

Disadvantages for Owners

Every boatbuilding material has its disadvantages, and fiberglass is no exception. Because of the high density of fiberglass, the interiors of fiberglass hulls are more prone to sweating than wooden ones. And most fiberglass hulls, especially those without a soft-core material between outer and inner layers of fiberglass, cannot match the solid sound of a wooden hull moving through the water. Methods for alleviating these disadvantages are discussed in later chapters of this book.

New fiberglass boats are expensive. Popular used boats, because of their high resale value, often cost almost as much as new ones. Often, however, they are better equipped, so this must be taken into consideration. At any rate, high initial cost must be considered a disadvantage.

Another possible disadvantage is that thousands of inadequate boats have been sold. This was to be expected when fiberglass boatbuilding was pioneered, but shoddy, even unsafe boats are still being produced. It seems to be easier to fool the boat buyer with fiberglass than with other materials.

Advantages for Manufacturers

Since manufacturers pay for the advertising in boating publications, they also, directly or indirectly, control the content of the articles therein. Thus, the public is given only one side of the story. There are many advantages of using fiberglass for the manufacturers that result in disadvantages for the purchaser.

The minimum boat represents one such disadvantage. A boat can be built many times stronger than is necessary for the job or function it is to serve. This is the safety margin or factor. With a wooden boat, it's generally impractical to vary the thickness of the planking, so it is common practice to cover everything with the thickness of the thickest layer required. Varying the thickness in a fiberglass boat is relatively easy; thus, the boat can be engineered. Unfortunately, more and more production boats are being built closer and closer to the minimum, that is, with a smaller safety factor.

Fiberglass boats are much more adaptable to assembly-line methods than wooden or metal ones. Larger manufacturers now have separate molding and assembly sections. Though only one boat at a time can be molded in the standard contact mold, the assembly can be done on a line basis.

A typical production problem is that some jobs are more time consuming than others. For example, hull-to-deck joints have long been a problem area. Manufacturers have looked for faster and less costly ways of doing this operation. Some of these methods have resulted in sacrifices in the basic integrity of the boats, yet these boats are advertised as superior. The same situation exists regarding sub moldings. These can improve the strength and appearance of a boat, but this is often a secondary consideration from the manufacturer's point of view.

Though some manufactured fiberglass boats start falling apart before they leave the factory, some quality boats are also available. Price is often, but not always, a guide to this. The reputation of the manufacturer is the best assurance of quality, but don't judge this by the amount of advertising, for a company pays to tell you what it wants you to hear. Advertising is likely to catch the newcomer; experienced boatmen are more difficult to fool.

Selecting and Buying a Boat

Important considerations are design, quality, price, reputation of the manufacturer, and resale value. The potential buyer should learn the basics of boat design. Important factors in boat design are seaworthiness, comfort, performance, and appearance.

Experience is needed to recognize quality in a fiberglass boat, as much of it is hidden from the casual eye. Always try to talk with other owners of the same model of boat. Since a boat is a large investment, a survey of a used boat makes good sense, and is well worth the cost.

Price often returns the potential buyer to reality, often making it necessary for him to settle for less than he originally had in mind. In many cases, however, the modifications shown in this book can make the boat closer to what the buyer wants.

Boat manufacturers have, in general, moved away from making custom boats and toward making many boats the same. This means that the boats are made for the average buyer. In practice, however, there doesn't seem to be any average buyer, and most buyers must settle for "not quite" what they want. Again, modifications may be the answer.

A final point in buying a boat is its resale value. Find out what used boats of the same model are selling for in comparison with new ones. A low resale value often means that the boat was unsatisfactory in one or more ways.

Extras

This is a big gimmick. In most cases, the "base" price of a boat bears little resemblance to the final cost. Added to the base price is commissioning and so-called "extra equipment," which often includes items such as sails, engines, bilge pumps, and running lights. I've seen cases where, when the final checks were written, the buyers paid more than twice the "base" price.

Used boats are often fairly well equipped, but even so there's generally room for improvement. By adding your own extra equipment, as detailed in this book, you not only avoid installation labor costs but also, in many cases, pay less for equipment, which can be purchased at discount prices.

2

FUNDAMENTAL TECHNIQUES

In many ways this is the most important chapter in this book. By learning a relatively few fundamental construction techniques, you can make hundreds of modifications on a stock fiberglass boat. Since quality workmanship depends on doing these fundamental techniques well, I have included in the text practice projects, many of which can be done with scrap materials, so that you can develop these skills and determine if you are ready to make similar modifications on your boat.

It's surprising how many people attempt their first fiberglassing in an actual modification on their boat, only to find that they do not have enough skill to do a good job. The first attempt at almost anything is likely to involve a lot of fumbling. A very important rule, then, is not to do the practicing on the boat, but rather to wait until you have the skill and confidence first.

WORKING WITH CURED FIBERGLASS

Cured fiberglass, such as the hull shell of the boat, can be drilled, sawed, filed, sanded, and polished. It's extremely difficult, however, to punch or shear and it cannot be hammered or bent into a new shape. The shaping was done in a mold with "wet" fiberglass.

To practice these fundamental skills, use scrap pieces of cured fiberglass. These can generally be obtained from fiberglass molding shops. If not, a sheet of fiberglass purchased at a hardware store can be used, although the quality is generally inferior. It will serve the purpose of practice as long as you keep in mind that the boat laminate is, hopefully, of better quality.

The tools needed are discussed with the job in which they are used. Metalworking tools should be used on cured fiberglass. Fiberglass is slightly abrasive and will dull tools quickly. It will often ruin woodworking tools.

Notice that the laminate is made of layers of fiberglass reinforcing materials (these are discussed in detail later in this chapter) and resin. On the side that was against the mold is a colored layer of resin – actually pigmented resin – called "gel coat." The fiberglass reinforcing material gives the laminate its strength. The resin layer is rather brittle. If you have a piece of scrap material from a fiberglass boat molding, you can use a sharp object to demonstrate these characteristics. Notice that the gel coat scratches rather easily. Try chipping the piece of fiberglass first on a flat side and then on one end. In most cases, you will find that resin alone chips rather easily, especially at the edges of the laminate. Understanding these characteristics is important in working with cured fiberglass.

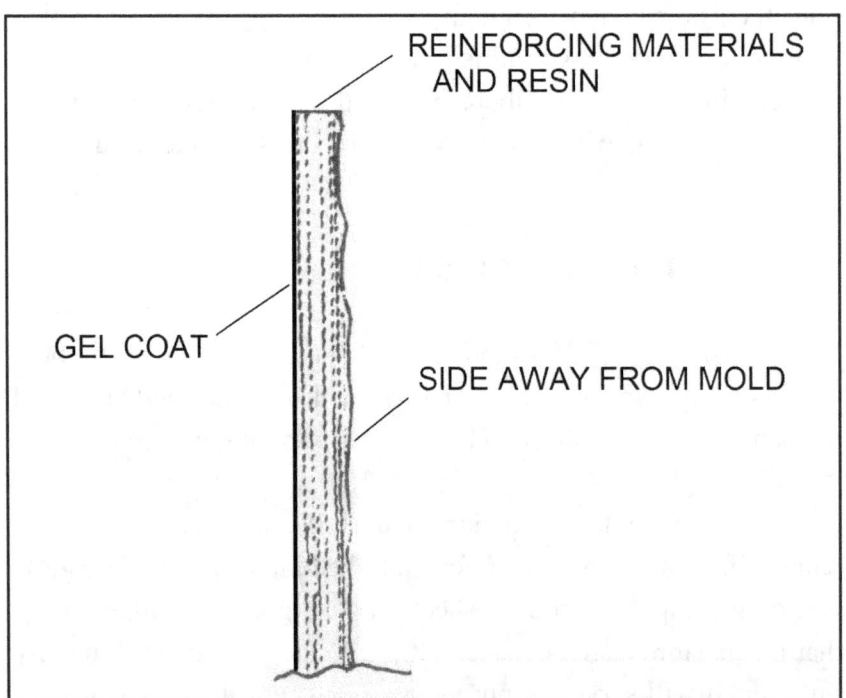

Fiberglass laminate.

Drilling

Many modifications require drilling cured fiberglass. Though a hand drill can be used, a portable power drill is much easier. When working with tools, whether hand or power, follow good safety practices. Always use appropriate safety devices. Always know what you are going to do before you do it. Always quit working when fatigued. Make sure the electrical circuits for all power tools are grounded properly.

As a general rule, holes should be drilled starting from the gel-coat side of the laminate. Drilling in this way, a section of the gel coat will not be chipped away when the bit goes through. To avoid damaging the gel coat around the location where the hole is to be drilled, drill a hole through a scrap piece of wood. Then place the wood over the fiberglass so that the hole in the wood lines up with the mark for the hole through the fiberglass. The hole in the wood will then serve as a guide.

Practice drilling various sizes of holes in scrap pieces of fiberglass. For small holes, standard metal bits can be used. For larger holes, a saw attachment with a metal cutting band can be

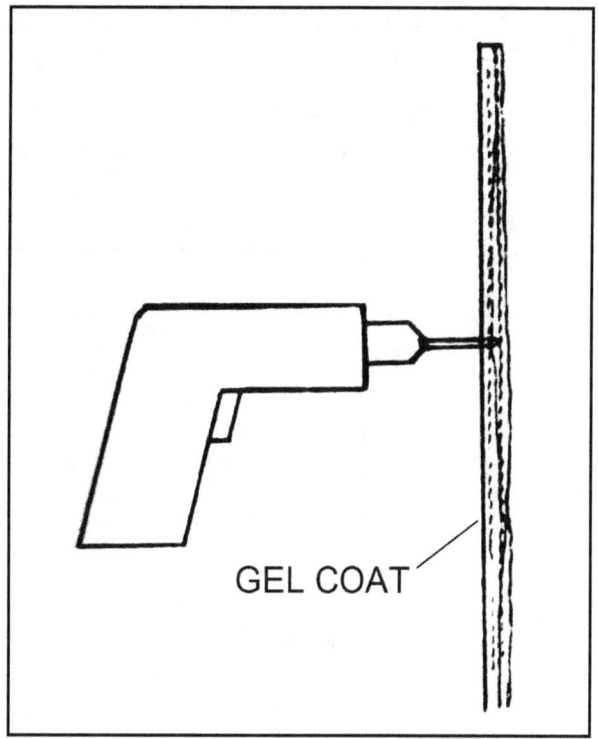

As a general rule, holes should be drilled starting from the gel-coat side of the laminate.

GEL COAT

used. These are readily available in 1/8-inch increments from ¾ to 2½ inches. A mandrel with a small bit is used to start the hole, and as a guide. Try drilling a hole from the wrong side so that the bit comes out on the gel-coat side. This will probably chip out pieces of gel coat around the hole, showing why this direction of drilling is generally avoided when working on a boat.

A common problem in drilling is not holding the drill at the proper angle, which is generally, but not always, at right angles to the surface being drilled. One simple way to drill holes at a desired angle is by using a hole drilled in a block of wood as a guide. Another way is to use a small drill press. A portable power drill can be attached to some of these. The drill press can then be clamped or blocked in place where the hole is to be drilled. Practice drilling holes in the scraps of fiberglass until a hole of the desired size can be drilled at the desired angle without damaging to any serious extent the fiberglass molding surrounding the hole.

Sawing

There are many modifications that require sawing cured fiberglass. Hand saws can be used, but an electric saber saw is easier. A metal cutting blade should be used for fiberglass. In most cases, it is best to face the gel-coat side of the laminate when sawing so that the blade guide will be against the gel coat.

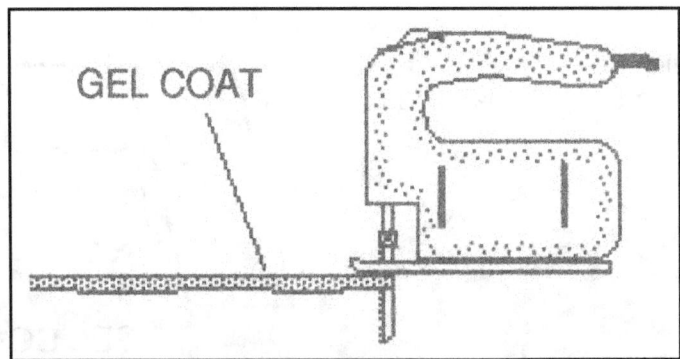

Sawing fiberglass with a saber saw.

Using scrap pieces of cured fiberglass, practice sawing. Try marking a pattern and then following the center of the line. Also, try drilling a hole and then cutting a circle or other pattern along a pre-drawn line. This operation is used in many modifications, such as making cutouts for windows and ports. In most cases, it's best to leave the line showing when making a cutout. In this way the opening can be filed to the final size.

Filing

Metal files can be used for filing cured fiberglass. The type of metal file to use depends on the particular job: a flat file for straight edges, and mill, round, and half-round files for various curved surfaces. Since fiberglass is fairly easy to file, a fine file can be used for many jobs. Course files can be used for rough filing and for filing down large areas.

When filing, it's generally best to face the gel-coat side of the laminate, as shown below. By filing in this way, chipping away pieces of gel coat can be kept to a minimum. This is especially important if the edge will show.

When filing cured fiberglass, it's best to face the gel-coat side of the laminate.

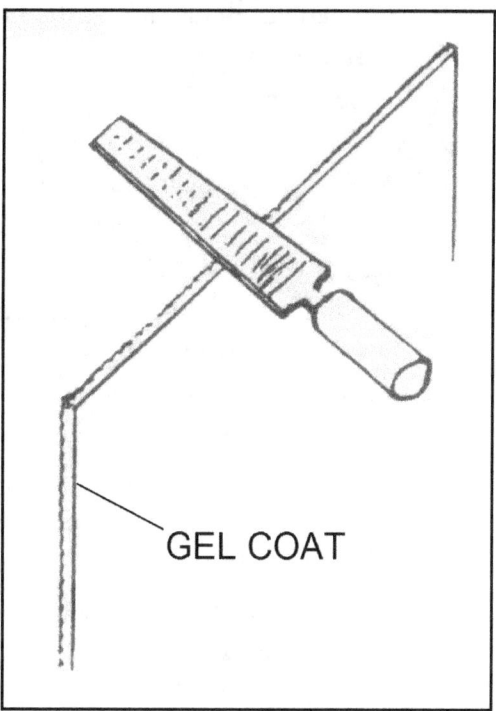

GEL COAT

If a flat edge is desired, the file should be held level and an even pressure applied. Practice this on scrap pieces of fiberglass. Also, try rounding off a square corner and both edges along one side. If the file becomes clogged, you can use acetone to clean it. Acetone is a solvent that is widely used in fiberglassing. Although many people use acetone to clean their hands, this is not recommended. Much safer is a barrier cream designed especially for working with fiberglass. Several brands are now on the market.

Sanding

Next, try sanding fiberglass. This can be done by hand or with a disk, belt, or orbital sander. For jobs involving limited sanding, power sanders are not necessary. However, if you plan extensive modifications requiring the sanding of large areas of fiberglass, you will probably want to invest in at least a disk sander. This is an extremely versatile sanding tool, but it takes considerable practice to develop the skill to use it effectively.

Abrasive papers, also called "sandpapers," are used extensively in fiberglassing. The most common types of papers are garnet, aluminum oxide, and silicon carbide. These come in various grades, four of which are recommended for experimenting: 50-grit for coarse work, 80-grit for medium work, 120-grit for fine work, and 220-grit for very fine work. In actual finishing work, paper as fine as 600-grit is sometimes used.

Some papers are suitable for "wet sanding," a technique that can be used when a very fine finish is required. For example, wet sanding with 220-grit paper can remove the scratches left from dry sanding with 120-grit paper. Wet sanding is perhaps superior to dry sanding for some finishing jobs on fiberglass and metal. One reason for this is that wet sanding seems to leave fewer scratches than dry sanding. Wet sanding should be done by hand, *never* with a power sander. Soak the paper in water and keep it wet while sanding.

A sanding block is an important tool for hand sanding. A wooden block can be used.

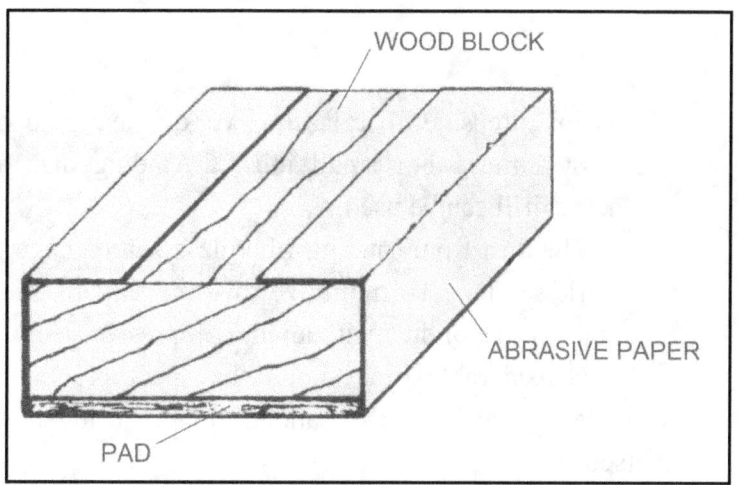

Sanding block.

Special sanding blocks with clamps for holding the paper in place can be purchased. A pad of sponge rubber, felt, or carpet is often used between the sanding block and paper.

Using scraps of fiberglass, practice hand sanding with various grades of paper with and without a sanding block. Try sanding both the gel-coat and the rough sides of the laminate. Also, practice sanding a sawed edge. Since the dust resulting from the sanding of fiberglass is extremely irritating to the skin, contact should be avoided as much as possible. Ordinary clothing is generally ineffective in keeping the dust from the skin. A rubber raincoat can help. A barrier cream designed especially for working with fiberglass should be applied to the hands and arms before sanding. Most important, avoid breathing the dust. A face mask should be worn, especially when using a power sander.

If you intend to use a power sander in making modifications, practice also with this tool. The easiest type to use is a vibrating model, as it poses little danger of over sanding.

A belt sander is generally faster, but more care must be taken to make certain that the sander is not held in one spot too long and that it is properly adjusted to avoid scarring.

The rotary-disk sander is probably the most versatile for sanding fiberglass if a variety of sanding and grinding operations are to be done with one tool. However, it takes considerable practice to avoid scarring and leaving swirls. A rubber cushion is often used between the hard disk and the paper when finishing work is done.

For extensive fiberglassing work, a heavy-duty disk sander is almost a must. For small jobs, a sanding disk attached to an electric drill can be used.

The dust from sanding fiberglass can damage the brushes on electric sanders. To minimize this, the sanders should frequently be blown free of dust with an air compressor.

Considerable time should be spent practicing with power sanders, especially disk sanders, before using them on the boat itself.

Polishing

One other important operation on cured fiberglass is polishing. This can be done by hand with a soft cloth and either polish or bugging compound. Buffing pads attached to rotary sanders can also be used.

The above five operations on cured fiberglass – drilling, sawing, filing, sanding, and polishing – form the basis for working with fiberglass laminates. Time spent learning these skills will not be wasted.

FIBERGLASSING

Many of the modification covered in this book require fiberglassing skills in operations such as bonding, reinforcing, embedding, patching, and in some cases molding.

The basic principle of what has come to be known as "fiberglassing" is reinforcing a plastic (resin) with glass fibers. The glass fibers are the load-carrying part of the combination. The resin holds the glass fibers in position. The combination of resin and glass fibers is typically referred to as "fiberglass."

Reinforcing Materials

Three basic types of fiberglass reinforcing materials commonly used are mat, cloth, and woven roving. The reinforcing materials are made from glass drawn out into fine fibers.

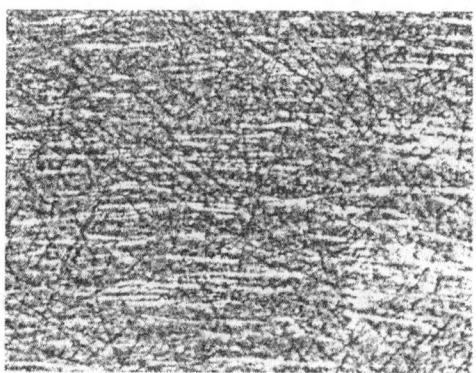

Fiberglass mat.

Mat is made up of a random pattern of chopped strands of glass fiber that are held together in a felt-like material by a bonding agent. Small quantities of mat are sold in individual packages; or it can be purchased by the foot or yard in rolls. The weight (for example, 3/4-ounce mat) refers to the approximate weight per square foot. Its main uses are to provide thickness, water tightness, and filler between layers of woven roving and sometimes cloth. A great deal of resin is required with mat, which results in a laminate that is not as strong as when cloth is used. On a weight basis, mat costs about half as much as cloth.

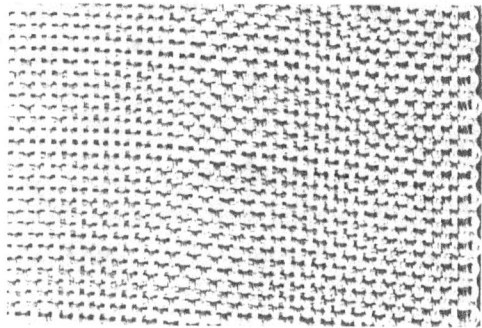

Fiberglass cloth.

Cloth is probably the fiberglass reinforcing material used most for modifying fiberglass boats. A plain weave pattern is usually used. Cloth is available in a variety of tape and standard widths with salvaged edges that prevent the cloth from coming unraveled at the edges. Once cut, however, this advantage is lost, so whenever possible cloth should be selected the proper width for the particular job.

Cloth is available in various weights per square yard, with 10 ounce and 12 ounce weights commonly used for boat construction work. Cloth gives a laminate that is approximately 40 percent stronger than a mat laminate of the same thickness. The disadvantages are the high cost and low shear strengths between layers.

Fiberglass woven roving.

Woven roving has a heavy, thick, basket weave. It has strengths approaching those of cloth at a price only slightly higher than mat. In laminates, mat is often used between layers of woven roving to help fill in the spaces in the weave pattern so that there will not be pockets of resin without reinforcing material. Like cloth, woven roving is available in various weights per square yard, with 24-ounce woven roving being commonly used in boat constructions.

Only glass-fiber reinforcing materials that have been treated to remove oils used in their manufacture are suitable.

Molded fiberglass hulls and other molded components used in manufactured fiberglass boats usually consist of a combination of reinforcing materials. When using fiberglass bonding strips and other fiberglassing techniques for modifying fiberglass boats, you can also apply combinations of reinforcing materials.

Glass fibers are also available as chopped strands and milled fibers, which can be mixed with resin to form putty.

A variety of non-fiberglass reinforcing materials, including polypropylene, polyester, acrylic, and carbon fiber, are now on the market. While generally more expensive than fiberglass reinforcing materials, these materials may have advantages for certain applications that offset the additional cost.

Resins

Two basic types of resin are used in boat construction work: polyester and epoxy. Polyester is less expensive and, thus, is used where suitable. Most fiberglass hulls and other boat moldings are made with polyester resin. It is also suitable for many bonding and reinforcing applications for modifying boats.

One formulation is used for bonding or laminating; another as a sanding or finishing layer. Bonding and laminating resins remain sticky even when they have set. Sanding or finishing resins contain wax and become hard, which allows them to be sanded. However, before another layer of fiberglass can be bonded to them, the wax must first be removed. Special wax removers or acetone can be used to remove the wax. A wax additive is available that allows bonding and laminating resin to be used as a sanding or finishing layer.

When you are ready to use polyester resin, a catalyst – also called "hardening agent" – must be added according to the directions for the particular brand you are using. At low temperatures, more catalyst is added, and at high temperatures, less. The amount of catalyst, assuming a constant temperature, determines the "pot life," which is the time you have to apply the resin before it begins to set up.

The catalyst is what causes the resin to set. Once the catalyst has been added, it's strictly a one-way reaction. Even sealing the resin in a container will not keep the resin liquid. An accelerator is also needed to cause the resin to set at room temperature, but this is usually mixed into the resin during manufacturing.

The most common catalyst used for polyester resin is methyl ethyl ketone peroxide (MEKP). It is important to remember that the catalyst must be added to the resin before it will set.

Epoxy resin generally gives a better bond than does polyester resin. However, it is more expensive than polyester resin. Epoxy resin is mixed with a curing agent or hardener prior to application. Epoxy resins generally take longer to cure than do polyester resins, which can make application more difficult.

Health and Safety Factors

It is important to consider health and safety factors before you start doing fiberglassing. Follow all health and safety procedures recommended by the manufacturers of the products used. Use good ventilation and avoid breathing the fumes from resins, catalysts, hardeners, acetone, and other chemicals used. Respirators, when required, should be of a type designed to give protection from the particular chemical being used. Change cartridges and otherwise maintain respirators according to the instructions of the manufacturer. Wash hands often with soap and water. Always know what precautions should be taken should any accident occur while working with a particular chemical. These are generally included on the containers.

If you develop dermatitis or any other adverse reactions from fiberglassing, seek medical advice.

Working Conditions

Working conditions can affect the quality of fiberglassing work. The best conditions involve controlled temperature, humidity, ventilation and cleanliness. You won't be able to control all of these things. Thus, you will usually have to do the fiberglassing with less than perfect conditions. Even so, you can still do the work satisfactorily.

The preferred temperature is about 70° F. By varying the amount of catalyst, however, it is possible to do satisfactory work from about 55° F to 85° F. Special resins are available for use in more extreme temperatures, but these are usually more expensive and difficult to use than ordinary resins.

Dry conditions are extremely important. Avoid working on damp and humid days. After fiberglassing, it's generally recommended that the area be kept dry for a day or two. By this time it will usually have cured to the point where it is virtually unaffected by water.

Good ventilation is extremely important when fiberglassing. This can be a real problem when working below decks. A fan mounted in a hatch or other opening for drawing air in can be helpful.

Fan mounted in forward hatch.

The work area should be as clean and free of dust as possible. Keeping areas clean around where you are fiberglassing is also important. Resin fingerprints and footprints can be difficult to remove from a gel coat

Areas that you want to keep clean can be covered with a sheet of polyethylene plastic or other suitable material. Use masking tape around the edges to hold it in place.

Tools for Fiberglassing

Only a few inexpensive tools are essential for learning basic fiberglassing techniques:

●Squeegee.

●Throw-away paintbrushes.

●A paint roller with extra covers.

●Cardboard or plastic cups or other suitable containers for resins. Make sure that they are suitable for resins and they are not waxed.

●Wood mixing sticks.

●Rags.

●Old scissors for cutting reinforcing materials.

●Laminating rollers.

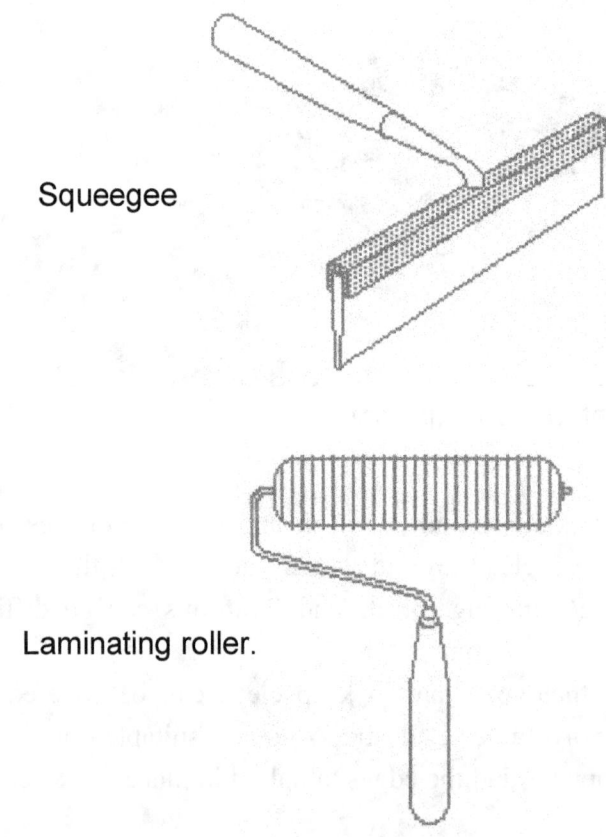

Squeegee

Laminating roller.

Basic Techniques

If you have never before done fiberglassing, I suggest that you learn the basic techniques before doing any modification on your boat.

Begin by organizing your work area, which should be as clean and dust-free as possible. For the first attempts at fiberglassing, I suggest that you use a piece of fiberglass cloth tape about 4 inches wide and 1 foot long. Begin by "dry" cutting the fiberglass tape with scissors to this length. This means that the cutting is done before any resin is applied to the reinforcing material. If you later do have to make a cut on reinforcing material that already has the wet resin applied, and this is sometimes necessary when you are doing actual boat work, you will need to clean the scissors with acetone to remove the resin before it has had a chance to set up.

For practice, the fiberglass tape will be laminated to a piece of scrap plywood. The surface of the plywood where the fiberglass is to be applied should be rough sanded and then thoroughly cleaned with acetone.

Have the fiberglass tape ready nearby. Spread it out on a piece of cardboard or plywood.

For this practice exercise, polyester sanding or finishing resin can be used. Later, you will want to use bonding or laminating resin if additional layers are used in the laminate. The sanding or finishing resin will be used for the final layer. Open the can of resin and mix it with a clean stick. This stick should be reserved for mixing un-catalyzed resin only.

Pour about an ounce of resin into a mixing cup. A cup with graded marks on the sides is helpful. You can also use a measuring scoop for transferring the resin from the container to the mixing cup. Once you have the desired amount of resin in the mixing cup, replace the lid on the resin container.

Next, add catalyst to the resin in the mixing cup. The catalyst is usually sold in containers that allow dispensing it by drops. Add the number of drops recommended by the manufacturer for the ounce of resin for the temperature you are working in. For the ounce of resin, 4 drops of catalyst should give a pot life (the time you will have to apply the resin before it starts to set up) of about 30 minutes at 75° F, and 8 drops about 15 minutes. Fifteen minutes should be enough time for the practice exercise.

Once the catalyst has been added, mix it into the resin with a small mixing stick.

Then use a 1/2-inch wide paintbrush to apply a thin layer of catalyzed resin to the area of the plywood where the fiberglass is to be attached. The fiberglass cloth tape is then placed over this and smoothed out. Apply an even coat of resin to the entire surface of the cloth tape. Work out any air bubbles with the brush. A squeegee or laminating roller can also be used for this.

If you have any resin left over, do not put it back in the container with the un-catalyzed resin. Discard it. If you need additional resin to complete the job, measure out and catalyze the required amount.

Use acetone to clean the brush before the resin has had a chance to harden. The fiberglass tape should form into a hard reinforced plastic material shortly after you have finished applying the resin.

Using the same basic techniques, practice applying mat and woven roving. Also try laminating more than one layer of the same and different reinforcing materials. Generally, the first layer will be allowed to set up before the next layer is applied.

Bonding Wood to Fiberglass

Some modifications involve bonding plywood parts (bulkheads, berth and cabinet sections, shelves, and so on) and other wood members to the hull and other fiberglass moldings. The bonding techniques should be practiced using scrap materials before trying them on an actual boat.

Plywood is frequently attached by bonding strips. Only marine or exterior grades of plywood should be used, never interior grade.

A good practice project is to bond a piece of plywood at right angles to a piece of fiberglass. Fiberglass bonding strips are then used to make the joint.

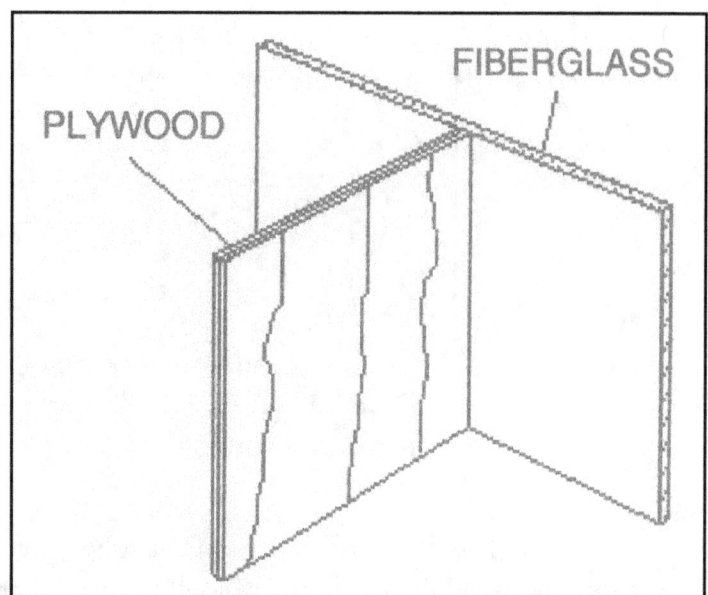

Practice project: bonding plywood to fiberglass.

The area of fiberglass where the wood is to be bonded must be clean. First remove any loose materials and then wash the area with acetone to remove wax and other contaminants. The plywood should be rough sanded in the areas where the fiberglass strips will attach.

The wood is then propped or clamped in place. In actual boat construction work, two or more layers of the same or different reinforcing materials are often used. For practice, cloth tape about 6 inches wide is recommended.

The next step is to cut the glass reinforcing material to the sizes needed. These should be arranged near the work area so that they are close at hand in the order that they will be used.

Measure out desired amount of polyester resin and add catalyst in the same manner as was done for the first practice project. Then, using a paintbrush, cover the area where the cloth will go with resin. This should be a thin, even layer. Next, lay the glass cloth in place. With the brush, cover it with a thin layer of resin. Then use the brush to smooth out the cloth as much as possible. A squeegee or laminating roller can be used to smooth out the cloth, work it into the corner, and work out any air bubbles. The most difficult area to smooth out will probably be the sharp corner. After the cloth has been wetted out on one side of the plywood, apply a second angle-bonding strip to the other side.

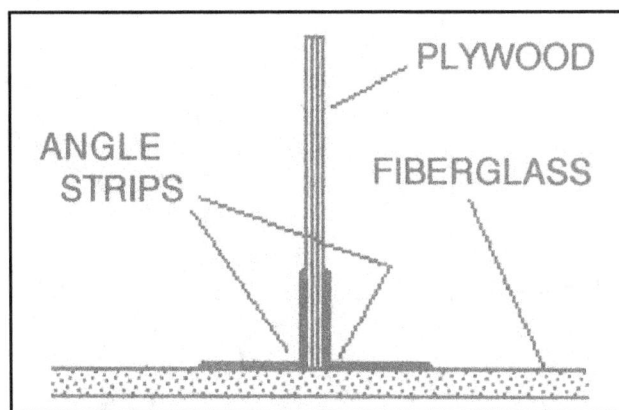

Basic angle bonds.

This is a basic method for applying an angle bond. In many cases, additional layers of cloth or other reinforcing material will be added, often in progressively wider strips.

A second method used for attaching plywood to fiberglass is basically the same except that fiberglass putty is applied to round the corner between the plywood and the fiberglass molding. The reason for this is that it is difficult to get fiberglass strips to bond properly in a sharp corner. After the plywood has been dry fitted and clamped or propped in place, the fiberglass putty is applied. Catalyst is first added to the putty. A wood dowel can be used as a drag for shaping the putty.

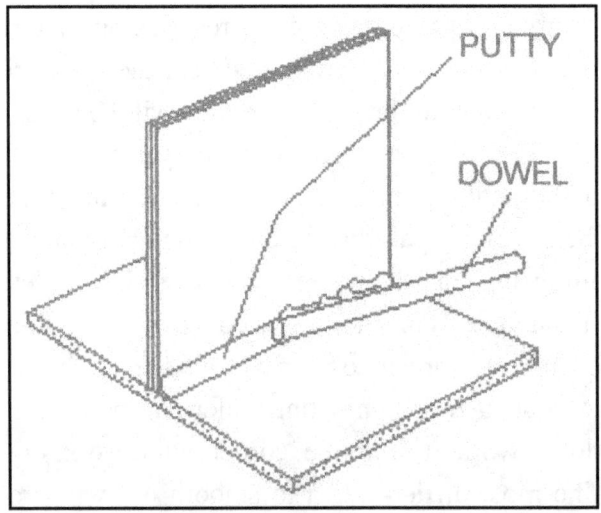

A wood dowel is used as a drag for shaping the putty.

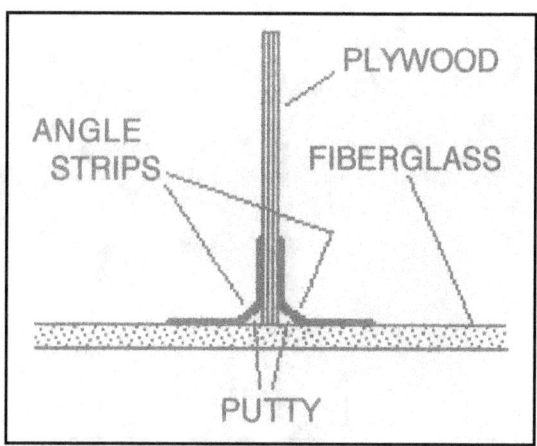

Putty filler is used to round corners
for angle bond.

A third method is to use a pad of polyurethane foam or other suitable material between the plywood and fiberglass molding. The pad allows rounding the corners and also helps to prevent a "hard spot." A "hard spot" results from the slight expansion and contraction of the fiberglass against a major structural member, such as a plywood bulkhead, which is almost non-shrinkable. A "hard spot" frequently appears as a surface disfigurement known as "print," which can be seen most readily in direct sunlight by sighting along the fiberglass molding.

The pads are typically about 1/4 to 3/8 inch thick. The plywood is cut smaller than a contact fit along the area to be bonded. The polyurethane-foam strip is generally about three times the width of the plywood. Some builders merely dry fit the polyurethane strips in place. I prefer to epoxy glue the strip to the fiberglass molding and the plywood edge. In either case, the wood member is clamped or propped in place. If epoxy glue is used, it should be allowed to set up before the bonding angles are added. This is especially important if polyester resin is to be used. Polyester and epoxy resins will bond to each other, but only after one has set up. Do not mix "wet" polyester and epoxy resins.

The next step is to shape the polyurethane with a rasp or other suitable tool to form a rounded angle between the shell and wood.

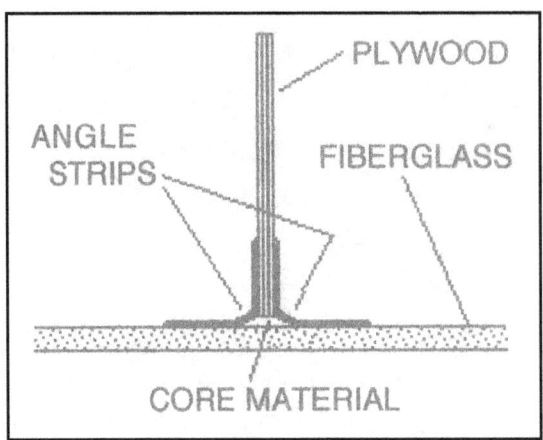

Angle bonds with core material.

Regardless of which of the above three methods is used, the fiberglass bonding strips are applied in the same way. On the boat itself the reinforcing material or materials, weights, and number of layers depend on the particular parts being attached.

Reinforcing Fiberglass

Reinforcing the fiberglass shell by adding additional layers of fiberglass to the laminate is a frequent modification procedure. A common application is strengthening an area of the shell for the attachment of a fitting. This is fairly easy when applied to a flat surface from above the surface. Vertical surfaces are usually more difficult. Overhead areas can be a considerable problem. Unfortunately, there are a number of modifications that require this. However, before doing this job, first see if there is an easier method of reinforcing the area.

The first step is to prepare the surface by removing paints and waxes from the areas where the fiberglass is to be applied. Rough sanding can also be done.

Next, cut the glass reinforcing material to size. The material used depends on the particular job. Generally, if mat is used, a layer of cloth or woven roving is placed over the last layer, with the cloth or woven roving extending an inch or two beyond the mat around the edges.

After a dry fit has been achieved, the glass reinforcing materials are put in place with catalyzed resin. In most cases, several layers of reinforcing material can be added at once. When working overhead, one layer can be allowed to set up before adding the next. Special bonding resins are available that make fiberglassing overhead easier.

When applying resin to glass reinforcing materials, remember that both too little and too much resin will lower the quality of the finished laminate. The amount is correct when the glass fibers are completely saturated without excess resin. Learning the correct amount generally involves considerable practice.

This reinforcing technique can be practiced by adding layers of fiberglass to a scrap sheet of cured fiberglass laminate.

Sheathing Wood with Fiberglass

Another important fiberglassing technique is sheathing wood. Although the trend is away from this, some manufactured boats with fiberglass hulls have wooden decks and cabin tops. In many cases, these are of plywood sheathed with one or more layers of fiberglass. In modification work, there are many places where it is desirable to sheath wood, especially plywood, with fiberglass.

Preparation of the surface is extremely important, as a good bond between fiberglass and wood is difficult to achieve. Begin by rough sanding the wood. It is also important that the wood be dry and free from oils and paints.

The technique for bonding a layer of fiberglass to the wood is similar to the method used for reinforcing an existing fiberglass laminate. As a practice project, add a layer of fiberglass cloth to one side of a small piece of plywood.

Finishing Techniques for Fiberglassing

In many cases the fiberglassing jobs require additional finishing work. The weave patterns of fiberglass cloth and woven roving can be filled with resin putty or surfacing compound. After adding the catalyst, the putty can be applied with a putty knife. Make the surface as smooth and even as possible. Small dents and holes can be filled in a similar manner. Larger defects, such as a sanded-off air bubble, may require glass reinforcing material.

After completing sanding, you should decide how the area is to be color-coated. The basic choices are gel coat and paint. Applying gel coat with a brush or roller is difficult, and an uneven finish often results. Spraying is better, but the cost of the equipment eliminates this method for most amateur builders. An alternative is gel coat in spray cans, such as Spray-Gel. These come in complete kits. The gel-coat resin is catalyzed as it leaves the spray can. Spray-Gel is available in a variety of colors, which will match many standard hull finishes, as well as in special colors.

Painting is another possibility. Epoxy and polyurethane paints are frequently used on fiberglass. The true epoxy paints must be catalyzed.

WORKING WITH WOOD

Many modifications to fiberglass boats, especially on interiors, involve working with wood. If you have had little or no experience working with wood, a course in basic woodworking is recommended. Many community colleges and adult education programs offer such a course.

Teak seems to be the most suitable wood to use on fiberglass boats, as it is most in keeping with the low-maintenance concept. The second choice is probably mahogany. These woods are expensive. If cost is a critical consideration, less expensive lumber such as Sitka spruce, white (clear) pine, and possibly even common spruce, fir and yellow pine can sometimes be used.

Plywood is used extensively in making modifications on fiberglass boats. Most common is plywood made of Douglas fir. I suggest using only exterior- or marine-grade plywood. In no case should any type of interior-grade be used, even if it is to be sheathed completely in fiberglass.

Shaping wood parts is frequently required in making boat modifications. This can be done with hand tools, but power tools are easier. A typical job involves cutting a piece of lumber or plywood to a certain pattern. Additional shaping, such as rounding edges, might be required. Drilling is another common job.

A number of factory-shaped components, such as handrails, cabinets, tables, racks, and shelves, are being manufactured. Although these items are rather expensive, they essentially guarantee a neat installation. These components are ideal for the amateur builder who does not have the desire or skill to make them himself. Typical manufactured components are of teak and mahogany, and are available finished or unfinished.

Many modifications require wood-to-wood joints. Glue with either screws or stronghold nails is commonly used. In a few cases, bolts are required. I suggest that the fasteners be bronze, stainless steel, or monel, and that they be of marine grade.

All parts should be dry-fitted first. If screws and plugs are to be used, a pilot bit that countersinks and counter bores in one operation will save time. After the holes have been drilled and the pieces completely fitted, the screws should be removed. A thin layer of glue is then applied to both surfaces. Though several types of glue are being used for boat work, I suggest that epoxy or waterproof resorcinol glues be used. Both require mixing two parts together.

Gluing is an important skill in boat modification work.

After the glue has been applied, the pieces are fitted and the screws put in place and tightened down. Plugs are available at marine stores in woods such as teak and mahogany; or you can use a plug cutter to shape your own. Dip the plug in glue, line the grain up, and tap the plug into place. Allow the glue to set up and then trim off the excess plug. Finish by sanding. Stronghold nails should be driven until they are almost flush with the surface. A nail set can be used to drive the heads slightly below the surface. Fill the holes with putty.

Moldings and corner pieces simplify construction and give a neat appearance. A corner post is frequently used when joining two pieces of plywood. The exposed plywood edge can be protected with a molding. An alternate method is to use a rabbeted corner post.

Corner-post construction.

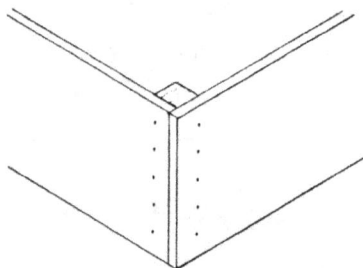

Rabbeted corner post.

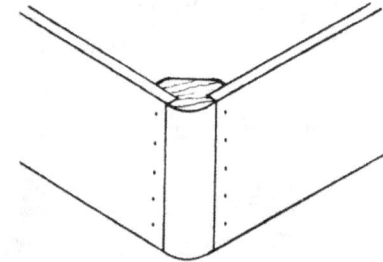

Cap molding.

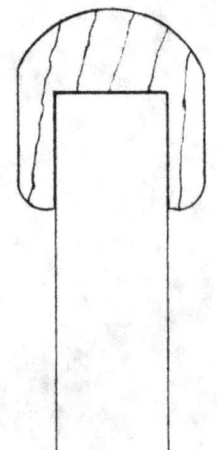

Molding.

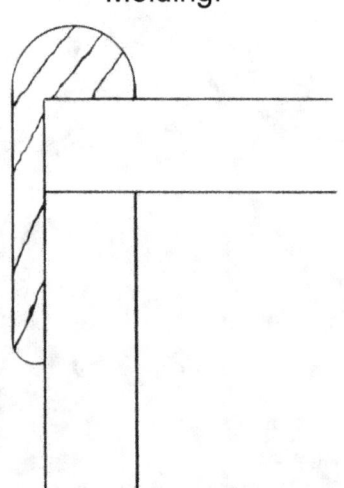

Corner construction with post and quarter-round molding.

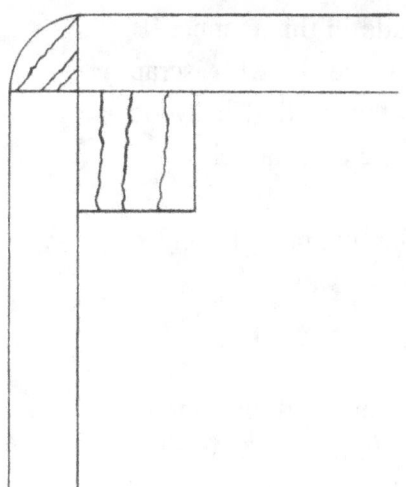

Built-up corner post.

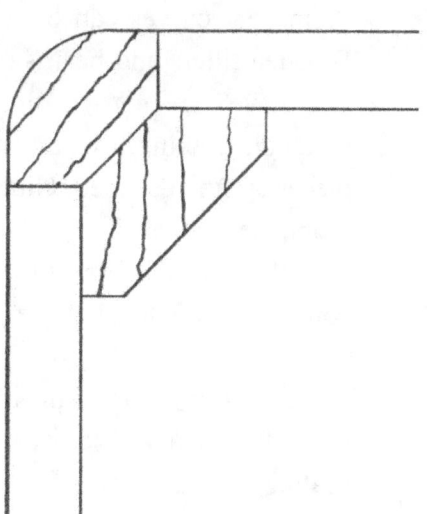

L-shaped molding used in corner construction.

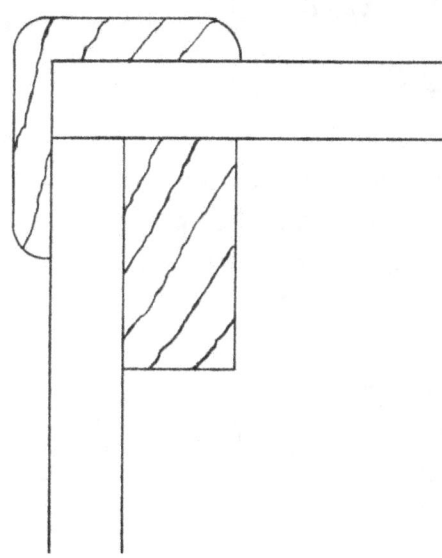

Laminating layers of wood in a clamp mold.

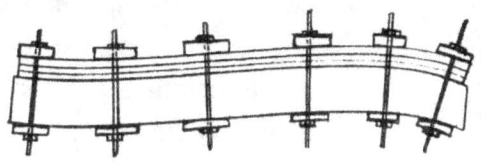

Method for constructing laminating clamps.

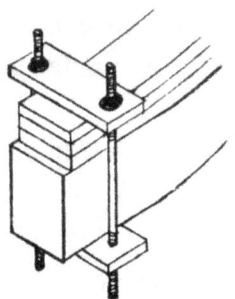

Another useful skill is laminating layers of wood together. Permanent curves can be made by laminating in a clamp mold. Wooden tillers and beams are frequently made in this manner. If you have never tried this technique, practice it with scrap material. Making a clamp mold is basic to this skill. The wood pieces are then shaped. Glue is applied, and the parts are clamped in the mold.

Plywood can be laminated in a similar manner. Rounded companionway hatches are sometimes made this way.

Finishing Techniques for Wood

After shaping comes sanding. Sometimes filling and/or sealing precede some of the sanding. A putty knife can be used to apply fillers and surfacing compounds. The basic sanding method is to gradually work down to finer grades of paper.

Wood surfaces can be finished in a variety of ways. Teak can be oiled with special preparations available at marine stores. I've tried several brands and have found differences among them, especially in how long the finish will last on an exterior surface. Other woods can be finished by painting or varnishing. Another possibility is applying laminated plastics to wood, which is covered in the next section of this chapter.

In painting and varnishing, the preparation of the surface is extremely important. In general, the surface should be clean and dry before applying the finish. Only quality paints and varnishes suitable for a marine environment should be used. Painting and varnishing are best done on dry days with the temperature between 60° and 85° F.

To insure a compatible system of solvents, sealers, primers, undercoats, and finishing coats, use products supplied by a single manufacturer and recommended by him for use together. Follow the manufacturer's directions for the products used.

Applying Laminated Plastics to Wood

Laminated plastics are frequently added to plywood surfaces such as table tops, counter tops, and bulkheads. The basic technique for adding a plastic laminate is as follows.

Special contact cement is used. Before applying the cement, make sure the surface is lean and smooth. The laminate should be cut to approximate size, allowing an eighth of an inch extra, all around. This excess can be filed off after the laminate has been cemented in place. Follow the directions for the brand of contact cement being used. Two coats are usually required on both the plastic and wood. Allow the cement to dry.

Brown wrapping paper is then placed over the cement area of the wood, and the laminate is placed over this. Position the laminate, slide the paper out, and tap the laminate in place with a rubber mallet or a hammer and a soft block of wood. Work from the center out to the edges.

Finish by filling the plastic to exact size and then sanding with a block and 120-grit abrasive paper.

If a counter or table top is to be covered on the sides as well as on the top, the sides are generally covered first. The plastic is then filed flush with the top. The top piece is then installed. When finished, the top piece will extend out to the side pieces.

OTHER USEFUL SKILLS

The skills and techniques detailed above in this chapter are basic to many modifications. Many additional skills, including metalworking, sewing, painting, upholstering, electrical wiring, plumbing, and rigging, are also useful. Many builders learn and improve their skills as they go alone.

PLANNING MODIFICATIONS

Make sure your modification plans are practical. Always consider the effect a particular change will have on the safety, performance, comfort, and appearance of your boat.

I recommend the following dimensions: standing headroom, 6 feet (minimum); sitting headroom, 36 inches above berth or seat (minimum); height of seats and berths (for sitting), 12 to 18 inches; height of counters with full headroom, 36 inches; height of counters with sitting headroom, 28 inches; width of aisles and doorways, 20 inches (minimum). Keep these in mind when planning modifications. Before constructing anything smaller, I suggest that you try out the size with a mock-up from scrap materials.

Unless you are sure you know what you are doing, I suggest that you have a competent naval architect draw up plans for major structural modifications, especially those that might affect the safety or performance of your craft.

Keep modifications as simple as possible, and make sure they are compatible with the rest of the boat in terms of quality, style, finish, and appearance.

The bases for a successful modification job are planning, organizing, and having the necessary tools, materials, and skills.

WHERE TO DO MODIFICATION WORK

This depends on the particular modification. Some jobs can be done with the boat in the water at dockside; others require a haul out. Since boatyard rates generally depend on how long the boat is out of the water, some modifications can be started with the boat in the water and then completed in a short time with a haul-out. An example of this type of modification is one that requires a through-hull fitting; everything except the through-hull fitting can probably be installed with the boat in the water, and, thus, a short haul-out is all that's needed. Modifications can often be planned so that jobs requiring a haul-out can be done when the boat is hauled for routine bottom cleaning and painting.

In areas where boats are hauled out for the winter, modification work can be done during this time. Trailer boats can often be modified on the trailer. Dinghies and other small boats can be worked on in garages.

The most extensive reconditioning that I did on a fiberglass boat was in the mid-1970s. The boat was an old 17-foot Dotline cabin cruiser that I had purchased in run-down and damaged condition at a low price, which included a boat trailer. I did the work with the boat on the trailer at the side of the trailer house I was living in at the time. This project took me over a year to complete. It involved repairing damage to the boat hull, designing

Dotline boat and trailer at side of trailer house at start of reconditioning.

Completely reconditioned Boat-Boat afloat in California Delta.

and installing my own interior accommodations, lots of fiberglassing, painting, and canvas and upholstery work, and rebuilding the boat trailer. A number of photos used to illustrate modifications later in this book are from this project.

After completing the modifications, I towed the boat, which I named Boat-Boat, to the California Delta and lived aboard it for over a year and a half. This experience is described in my book, *Living Afloat: My Ten Years of Living aboard Small Boats*. This book is available from Amazon.com in both printed and Kindle e-book formats. For more information, go to:

http://www.amazon.com/author/jackwileypublications.

3

THE BASIC SHELL

A first consideration for any fiberglass boat you want to modify is the basic shell of the boat. You will want to take a look at how the boat was designed and constructed in the first place and its present condition.

Deficiencies you will want to look for range from shoddy and incomplete bonding of stiffeners to the shell, to serious structural design problems, such as inadequate hull thickness and stiffening. The purpose of this chapter is to show you how to recognize and correct these flaws.

Before undertaking any corrections or modifications on your boat make a thorough examination of the basic shell, stiffeners, and bonding work. You will need good lighting for this. Areas that are hard to reach and see are likely spots for sloppy and defective construction.

You should be able to identify existing defects and shortcomings in your boat. More important, you can determine the feasibility of correcting them. This will depend not only on the problems, but also on your skill and experience in doing the work necessary to correct them.

Fiberglass Bonds

Most fiberglass boats have fiberglass strap and angle bonds, with or without mechanical fasteners, for attaching separate moldings together and attaching components to the shell. In small, open boats these bonds are commonly used to attach seats, gunwales, and buoyancy chambers to the hull. These components, along with hull shape and thickness, provide structural integrity.

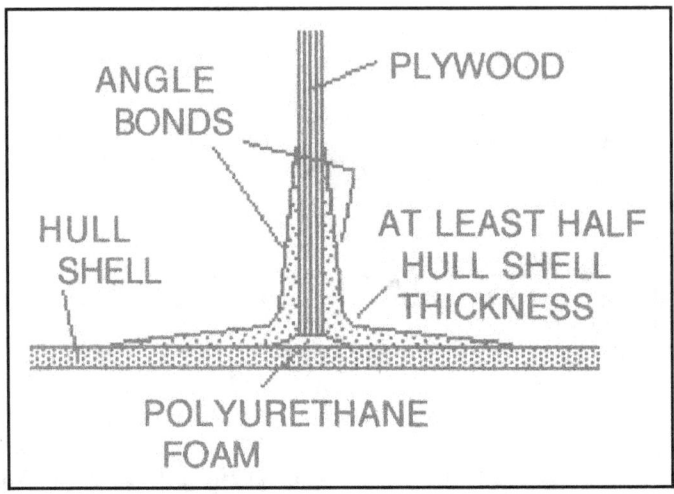

Adequate thickness and taper for angle bonds.

On larger boats, decks and cabin tops strengthen the hull by forming an enclosed shell. Additional support is provided by bulkheads and other stiffeners inside the shell. In boats with cruising accommodations, components such as built-in berths and lockers generally serve, in addition to their primary functions, to support the shell. Fiberglass bonds are used extensively to attach these components together and to the shell.

A recent manufacturing trend is to bond separately molded interior components, and sometimes complete one-piece hull liners, to the hull shell. These are designed to serve, in addition to other purposes, as stiffeners. The hull liner is general dropped into the hull while it is still wet in the mold. Fiberglass bonding straps and angles are commonly added.

Frequent bonding defects include air bubbles and fiberglass reinforcing material that was not properly wetted out or shaped. To correct these flaws, first scrape and sand away the defective areas. Small bubble areas and indentations can be filled with resin putty; larger defects can be filled with one or more layers of fiberglass mat and resin. If the area shows, a layer of glass cloth can be added. If possible, the cloth should overlap the original bonding strap or angle by at least 1 inch. Finish the job by sanding and, if desired, applying paint or gel coat.

Major Bulkheads

A critical structural defect in many stock fiberglass boats is inadequate thickness and/or taper of the bonding angles on major bulkheads.

Hard spots created by wooden bulkheads can lead to serious consequences. The problem can be largely eliminated by placing a soft core material between the bulkhead and shell before bonding. Proper shaping and tapering of the bonds are also important in preventing serious hard spots, as this will distribute the stresses over a larger area of the shell.

One way to check for hard spots on an existing fiberglass boat is to sight along the hull, which should be well polished, in direct sunlight. Hard spots will show up as "print" – a bulge in the surface of the fiberglass shell. For example, a marked bulge where a major bulkhead is installed indicates a hard spot. In serious cases, hull cracks can be seen along the edge of the hard spots. For this, major repair skill is required.

If marked print of a less serious nature is present and no core material was used between the bulkhead and the inside of the shell, you should decide whether or not the problem is serious enough to warrant adding a core material. This can be a considerable undertaking on a finished boat. It would have been a minor job during manufacturing.

Even if a core material is added at this point, the hull distortion may remain. However, the core material can make the hard spot less serious and thus prevent further trouble.

In order to add core material between a previously mounted bulkhead and the shell, it's necessary to remove the old bonding angles and a section from the wooden bulkhead. Both of these jobs are difficult. To prevent changing the basic hull shape, the work should be done in sections, generally a foot or less at a time. This means removing the old bond from the area, cutting away a section of the bulkhead, adding and shaping a soft-core material and bonding new angles in place. This new section should be allowed to set up before starting on the next area.

If a portable power saw is used, extreme care must be taken to prevent cutting into the shell. Use appropriate guides. The thickness of the section cut from the bulkhead should be equal to approximately half the thickness of the bulkhead. The soft-core material should be slightly thicker than the section cutaway from the bulkhead, and should be approximately three times as wide as the thickness of the bulkhead.

Though several different soft-core materials can be used, I recommend pre-foamed strips of polyurethane. After they are in place, they can be shaped with a rasp to form a gradual angle so that there will not be a sharp corner in the fiberglass bond. The contact areas of the shell and bulkhead should be coated with wet catalyzed resin before forcing the polyurethane strip in place. The bonding angles are then laminated in place.

If no core material is to be added, but the existing bonds are inadequate, proceed as follows: First, sand and roughen the areas to be corrected. If a sharp corner is present in the existing bond, you can add resin putty to form a smooth, gradual angle before you add additional layers of fiberglass. Then decide on the number of layers and their placement that will be needed to give the correct taper and thickness. Generally, mat with a layer of glass cloth on top is used. If the appearance of an area is not important, woven roving can be used in place of cloth.

When installing a bulkhead, generally apply the narrowest layer first, then the next widest, and so on. However, if a couple of wide pieces were bonded on with no regard for taper, you can add narrower layers on top to make the correct taper. Finish the lamination with a layer of cloth that extends over everything, with an additional overlap of approximately 1 inch on each side.

Some fiberglass boats have bulkheads that were bonded on one side only. By adding an angle bond on the other side, the strength of the bond can be increased and the stress spread over a larger area of the hull.

Bulkheads can also be shaped and added. The purpose might be to add a false transom in a runabout, to make watertight compartments, or to add a major stiffening member. Regardless of the purpose of the bulkhead, I recommend that it be installed with a strip of polyurethane between it and the shell. Most bulkheads are made from marine plywood.

Other Structural Members

In addition to major bulkheads, most larger fiberglass boats have several other transverse, longitudinal, and sometimes angled, structural members bonded to the shell. Some of these may serve other functions in addition to structure. For example, a sheet of plywood or a fiberglass molding may serve as a seat top as well as a longitudinal stiffener. On many boats, these bonds consist of a couple of layers of fiberglass slapped in place with no regard for taper or thickness. In many cases, no soft-core material is used between wooden components and the shell.

If no serious problem has developed as a result of the wood-to-shell contact, it probably isn't worth the trouble to add a soft-core material. However, it is frequently possible to improve the fiberglass bond without too much trouble. The same procedure used on bulkheads can be applied here, except that if the member is of minor structural importance, the thickness of the bonds need not be as great.

Angle between Transom and Hull

Correction in taper and thickness of the angle bonds between the hull and transom may be necessary, especially if the hull and transom are separate pieces. But even if the hull and transom are part of a single fiberglass molding, a plywood or soft-core reinforcement is often bonded in. In many cases, the taper and thickness of the bond can be improved. If a sharp angle was used on the original bonding, resin putty can be used to shape a smooth, gradual curve. A round dowel can be used to shape the putty. A laminate of mat, topped by a layer of glass cloth, can then be added.

Additional reinforcement may be required for mounting rudders or outboard motors. These jobs are detailed in later chapters.

Hull-to-Deck Joints

Another major problem area in stock fiberglass boats is the hull-to-deck joint. A number of chemical and/or mechanical methods are used to secure the two shell parts together. In an attempt to speed production, some manufacturers have resorted to mechanical methods where the shell parts are not in direct contact. Often, a metal extrusion is used. The two shell parts fit in notches, and bolts or rivets are used to hold the fiberglass moldings in place.

Problems with hull-to-deck joints range from annoying leaks to serious hull-deck separations. Three basic considerations from the boat owner's point of view are strength, water tightness, and appearance. On an existing boat, first determine the method used to join the hull to the deck.

Deficient extrusion joints that are found where the shell parts do not form a butt or lap joint are generally difficult to remedy.

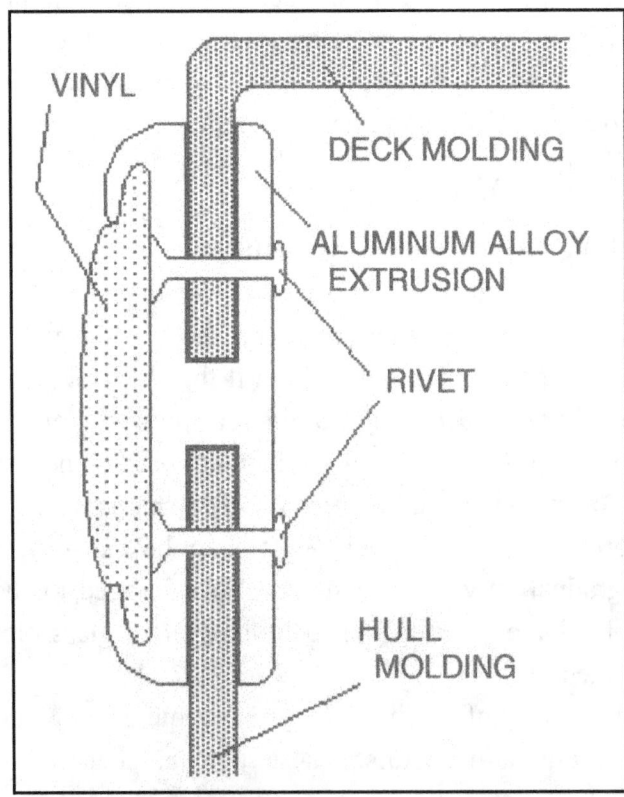

Hull-to-deck joint using an aluminum alloy extrusion.

Bedding compounds are difficult to apply after the joint has been made. Bonding across the metal extrusions on the inside of the boat might be done as a last resort. However, this is a major job, for which it is probably best to use epoxy resin. To be effective, the laminate must form a bridge that does not depend on the bond to the metal for strength.

Lap joints that are mechanically fastened without a fiberglass bond are generally easier to correct. A fiberglass bonding strap or angle can be added inside the hull. In order not to have to depend on the mechanical fastening, the angle or butt strap must be approximately as thick as the shell. Also, it is important to taper the bond away from the joint so that a serious hard spot will not be created in the shell. If the joint has failed previously, it may help to add a fiberglass bond on the outside of the shell as well. This makes a much stronger joint than if there is a bond only on the inside of the hull. It is difficult to finish an outside bond to a neat appearance, but sometimes it can be covered with a rail.

For hull-to-deck joints that were chemically bonded previously, with or without mechanical fasteners, you should check to see if the bond has been properly applied and if it is adequately thick and tapered. If not, the same corrections used on mechanically fastened joints should be made, as necessary.

Fiberglass reinforcing strap added to a mechanically fastened hull-to-deck lap joint.

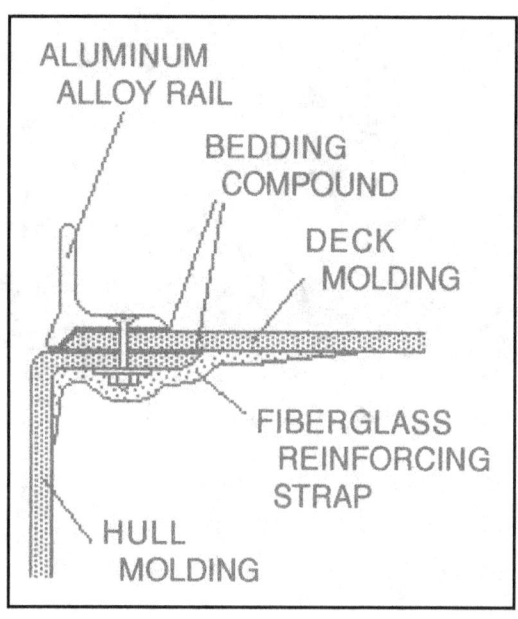

ALUMINUM ALLOY RAIL

BEDDING COMPOUND

DECK MOLDING

FIBERGLASS REINFORCING STRAP

HULL MOLDING

For minor leaks, epoxy sealing compounds are often effective. Try them before undertaking any of the more drastic measures.

Most decks and cabin tops are molded in one piece. However, if the cabin top is a separate molding, the same methods used on hull-to-deck joints can be used to correct deficiencies.

Repairing Hull Damage

On the Dotline power cruiser I reconditioned, there was actual hull damage at the center line near the bow. I repaired this damage by sanding off and tapering the edges of the damaged areas and fiberglassing a patch to these areas. The damage did not go all the way through the hull, so no backing patch was needed. In more serious cases, a fiberglass backing patch inside the hull would be required.

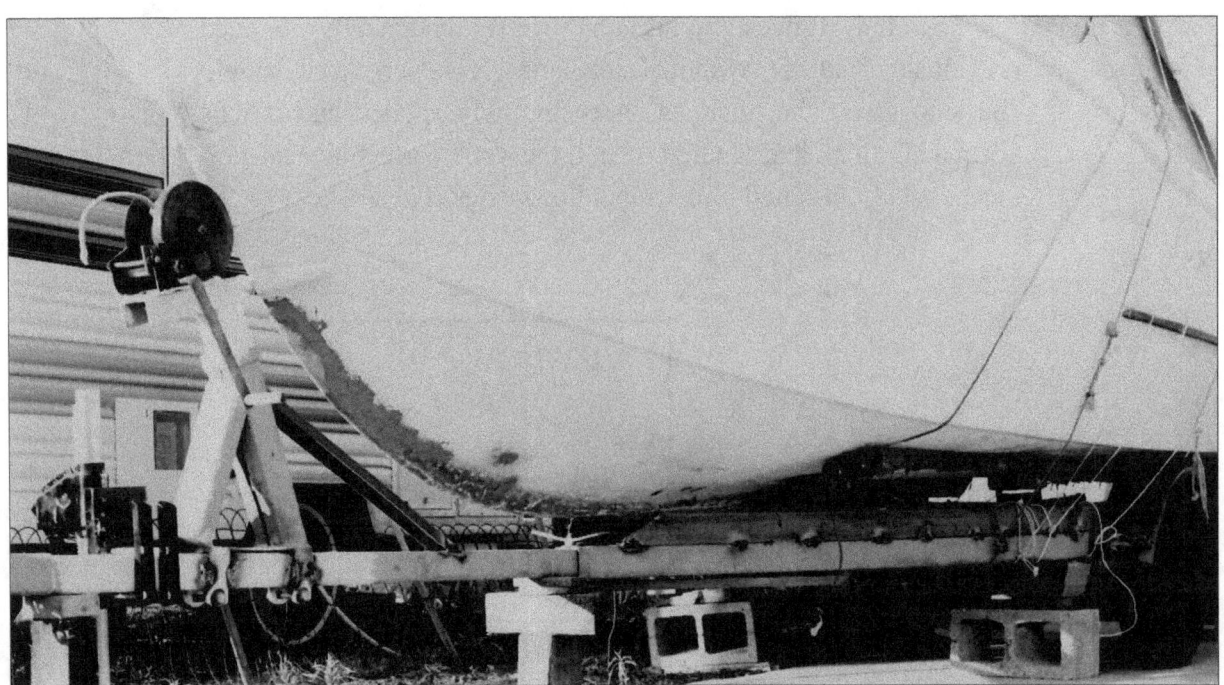

Fiberglass repair made on Dotline hull.

Adding Shell Thickness

Most stock manufactured boats have ample shell thickness. However, in some cases, added layers of fiberglass might be desirable. This is especially true if the boat is being modified for uses involving stress loadings beyond those for which the boat was designed. It should be understood that thickening the shell beyond a certain point, whether the hull is of single-skin or sandwich construction, is one of the least effective ways of adding strength. For both single-skin and sandwich hulls, molded stiffeners are often more effective.

Increasing hull thickness by adding layers of fiberglass to the laminate is best done before the hull leaves the mold. To add thickness in this manner to a completed boat can be involved and difficult. If additional layers are added, they are generally angled into bulkheads and other members attached to the hull.

Weight is another important consideration. On a small dinghy, one additional layer of glass cloth might add enough weight to decrease performance significantly. On larger boats the added weight is generally less critical.

Sandwich Construction

Adding a soft-core material to a single-skin shell and then one or more layers of fiberglass over this is generally much more effective than adding fiberglass alone. The sandwich is best formed while the shell parts are in the mold. However, in some cases, a sandwich core can be added to a completed boat. In addition to increasing strength and rigidity, it can give the boat a more solid sound. This is especially true when the core is added to the bow area. It's also an effective method of beefing up a deck.

Installation involves sanding and roughening all shell areas where the core material is to be added. Generally, pre-foamed sheets of polyurethane are used as a core on completed boats. The polyurethane is epoxy-bonded to the shell. Three-eighths– to one-half-inch thickness is about right for most jobs. This provides an effective sandwich, yet it's thin enough to be easily formed to the shell contour.

The fiberglass skin added over the core material must be thick enough to withstand local impact, abrasions, and handling. It is also important to consider weight, especially on a small boat. The added inside layers can rejoin the hull or deck, or they can be bonded into bulkheads and other members attached to the shell.

Applying fiberglass overhead, such as to the deck or cabin top from inside the boat, presents additional problems. A reinforcing material that is easy to wet out is helpful. A thixotropic resin, which does not tend to run on its own, also makes the job easier.

Molded Stiffeners

Molded stiffeners are ideal for general stiffening of the shell. They can be fabricated in place over a soft-core material. Considering the small amount of weight added, these molded stiffeners are extremely effective.

Although many lightweight core materials have been used, strips of pre-foamed polyurethane are recommended. Rectangular shaped foamed pieces will serve most purposes. However, if desired, half-round and triangular shapes can be used.

For maximum strength and rigidity, use woven roving and laminate to a thickness approximately equal to that of the shell. The taper should be the same as that of the bulkheads. Try to end all stiffeners on some other member already attached to the shell, or completely encircle the inside of the shell. It may be undesirable, however, to reduce ceiling height; in this case, the

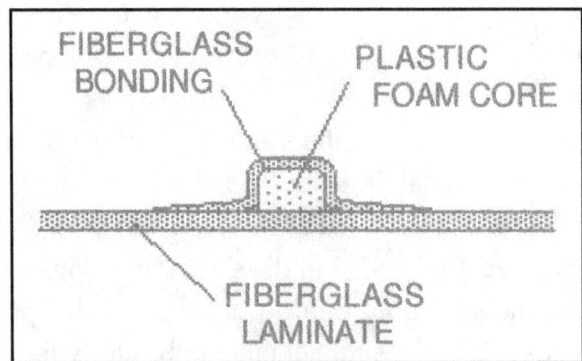

A molded stiffener.

stiffener can end at the hull-to-deck joint, or it can be ended by tapering it into the shell.

If the molded stiffener passes through the bilge, limber holes should be added. The holes should be lined with fiberglass.

Hull Liners

Both partial and complete molded-fiberglass hull liners are used on some manufactured fiberglass boats. Bonding straps and angles are commonly used between the main shell and these liners. The liners are generally designed to serve, in addition to other purposes, as stiffeners. Thus, it is a good idea to check these bonds for adequate thickness and correct taper and to make the necessary modifications.

Buoyancy Compartments

In addition to providing positive buoyancy in case of swamping or capsizing, buoyancy compartments often serve as important stiffening members. The bonds used to fasten these compartments to the shell should be checked for adequate thickness and taper, and corrected as required.

Buoyancy compartments can be added. On some boats, however, achieving positive buoyancy is in this manner is impractical, as it requires filling in too much of the interior space.

Flotation materials are available that are pre-foamed (in blocks and sheets) or that can be poured or sprayed in place. Polyurethane foam in spray cans is commonly used for small jobs.

In constructing buoyancy compartments, it is important to taper all fiberglass bonds into the shell. The skin added over the core material should be thick enough to withstand local impact, abrasions, and handling.

Basic Shell Integrity

A sound shell forms the basis for all of the other modifications shown in this book; without it, a really good boat is not possible. Thus, the basic shell integrity has been considered first. It's the basic foundation for a boat. I hope that manufacturing standards will improve to the point where an adequate shell and stiffeners are used on all boats produced. To date, however, this has not always been the case.

4

KEELS AND CENTERBOARDS

Many types of keels and centerboards, both ballasted and non-ballasted, have been used on fiberglass boats. Major modifications, such as changes in keel shape and size, generally present complicated design problems, which require the services of a naval architect. Such major modifications, which involve advanced boatbuilding skills, are beyond the scope of this book.

The types of modifications considered in this chapter involve correcting sloppy and inadequate workmanship, strengthening keels and centerboards, and making minor repairs, such as sealing around leaking keel bolts.

Internal Ballast

Properly installed internal ballast can eliminate many problems associated with external ballast, such as electrolysis and corrosion. Most ballasted powerboats and larger sailboats, with the exception of fin keels, are now using this method. Some fin-keel boats also use internal ballast.

Regardless of what metal or other material is used for the ballast, it should be sealed over, and water should be able to flow smoothly and easily from all areas of the bilge to the bilge sump. Typical improvements are filling in low areas where water does not drain properly, correcting sloppy and inadequate fiberglassing, and, in some cases, reinforcing the keel box above the ballast with additional layers of fiberglass.

In some boats, the limber holes through bulkheads and other members that extend across the bilge area are too small for adequate draining. These can be enlarged. Holes through fiberglass members should be lined with a layer of fiberglass.

External Ballast

External ballast is generally through-bolted to a reinforced area of the hull or to a keel box, which forms part of the keel. With external ballast, the weight can be concentrated at a lower point. However, several problems are frequently encountered, such as electrolysis, corrosion, and leaking around keel bolts.

Leaking around keel bolts can sometimes be corrected by removing the keel nuts and washers one at a time, applying bedding compound, and replacing the washers and nuts. This can generally be done with the boat in the water.

External modifications require a haul-out. A number of epoxy compounds that provide effective protection for metal keels are now on the market. In most cases, the epoxy is applied to bare metal. Sand blasting is recommended, but a power brush can also be used. In any case, preparation of the surface is necessary to achieve a good bond. The manufacturer's instructions for applying epoxy should be followed carefully. A primer may be necessary before applying antifouling paint.

A second method of keel protection is to use epoxy and fiberglass cloth to form a complete shell over the metal keel. This shell is generally extended beyond the metal to overlap the fiberglass shell. Several layers are required over the keel-to-shell joint.

The effect on performance of covering over a keel depends on many things. For a cruising boat, the advantages in keel protection generally outweigh the slight loss in performance. For a racing sailboat, this method is generally not used.

In some cases, replacing a fin keel is desirable. A new keel can generally be ordered from the manufacturer, or can be cast from iron or lead at a foundry. There are many problems, however, in removing the old keel and fitting the new one, so this method should be considered only as a last resort. Moving, fitting, and installing a keel will certainly challenge the skill and patience of most amateur builders.

After a dry fit has been achieved, either a dry or wet method is used to attach the keel or external ballast to the hull. The dry

Installing a fin keel on a fiberglass boat.

method generally makes use of a flexible bedding compound or a gasket. In the wet method, the metal is chemically bonded to the fiberglass. Epoxy is generally used, often with a layer of fiberglass mat. In both the wet and dry methods, the keel bolts are the main load carriers; the bedding compound, gasket, or chemical bond is primarily for sealing.

Non-ballasted Keels

Many powerboats have non-ballasted keels. Possible modifications include improving drainage to the bilge sump and strengthening the keel. These jobs are the same as were described previously in the section on internal ballast. If the keel area has been severely weakened from wear, it may be necessary to add reinforcing layers of fiberglass to the outside of the hull.

Swing Keels

A number of trailerable sailboats have swing keels. Some of these have proved to be satisfactory; others have been a source of problems, such as jamming and having too much play. The manufacturer of the boat should be consulted for methods of correcting these difficulties.

Sometimes, owners want to convert swing keels to fixed keels. In some cases, the manufacturers offer conversion kits. If they don't, the services of a naval architect are recommended, for the placement and amount of ballast is a design problem beyond the scope of this book.

Centerboards

The two main variations are the simple drop-in dagger board or pivoted centerboard, and the combination keel-centerboard. Some centerboards are ballasted or made from heavy metal. In the first variation, a centerboard trunk is typically bonded to the inside of the hull shell. Such constructions are typically found on small day sailers. Common modifications include strengthening the bond between the centerboard case and the hull and improving the fit of the centerboard.

The trunk can be reinforced by adding additional layers of fiberglass mat and woven roving. Where appearance is important, a final layer of cloth can be used.

A centerboard that does not fit properly can be trimmed down or thickened, as required. Both wood and fiberglass centerboards can be made thicker by adding layers of fiberglass.

A combination keel-centerboard allows shallow draft with the board up. On some designs the boards tend to jam, up or down. There are many reasons for this and other problems. The first step in making corrections is to identify the cause of the trouble. In some cases, the modifications are easy, but in others they are all but impossible. Occasionally, the manufacturer of the boat can be of help.

5

COCKPITS

Self-bailing cockpits are now standard on most manufactured powerboats and sailboats intended for use in unprotected waters. Their purpose is to achieve watertight integrity. Self-bailing cockpits are also desirable for closed boats that are to be used only on inland waters, especially if the boats are to be left unattended in the water. This way, bailing is not necessary after each rainstorm.

The principle of the self-bailing cockpit is simple. The cockpit is a well with the lowest point, the floor, above the water line. Drains are provided from the cockpit floor to a lower point outside the hull. However, to achieve watertight integrity, additional factors must be considered. Large-diameter scupper pipes or hoses should be provided so that the cockpit will drain rapidly. In general, they should be located at the low point of the cockpit well. The volume of the cockpit to the level where it will drain over the coamings should be small so that when the cockpit is filled with water, freeboard and stability will not be reduced to a dangerous degree. The water tightness of the cockpit should be such that even when it is filled to the highest level possible, no water will go below.

Modifying a Cockpit

I begin with the reconditioning and modifying I did in the cockpit area of a Dotline power cruiser. As in all reconditioning and modifying projects, you start with the way it is. What I started with was a complete mess.

I began by stripping out everything I did not intend to use or was in the way in areas where I intended to work.

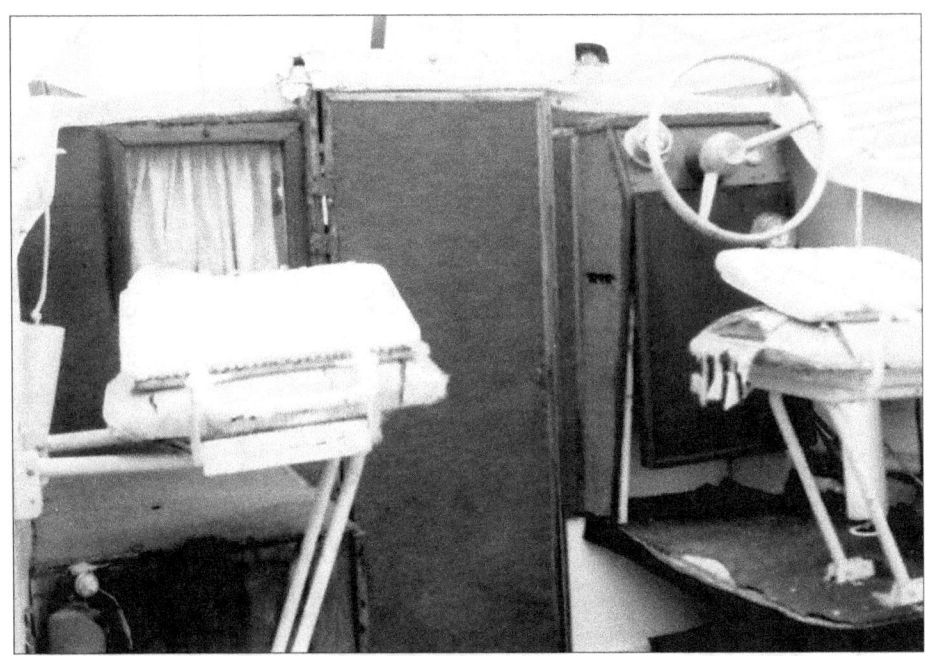

Cockpit area of Dotline at start of reconditioning project.

Cockpit area of Dotline stripped down for reconditioning.

I next did fiberglassing work on the cockpit floor, which was plywood sheathed with a thick layer of fiberglass. At the doorway to the cabin, I installed a piece of wood and then fiberglassed over this.

Applying fiberglass in cockpit area of Dotline.

I then installed a new steering station and seat of my own design. I tried to keep everything as neat and simple as possible. I used a dresser stand and a seat that I found at a thrift store. I sanded and painted the wood parts of the construction.

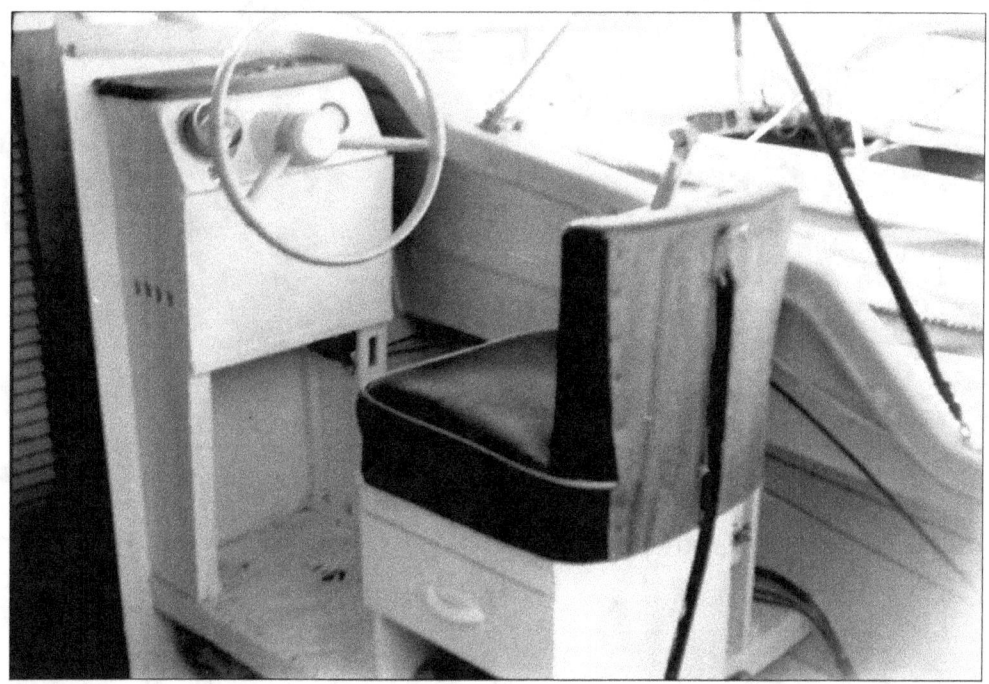

Completed steering station on Dotline.

Other modifications I made to the cockpit area included pads and hold down straps for carrying a fiberglass sailing dinghy in the cockpit and a place for carrying the mast and oars in the cockpit with a hole into the cabin area for the mast. A large gas tank for the outboard motor was mounted at the side of the cockpit. A frame was built over it as part of the dinghy mounting arrangement.

The dinghy mounting rack was used mainly for carrying the dinghy while the boat was on a trailer. However, it could also be used while the boat was afloat.

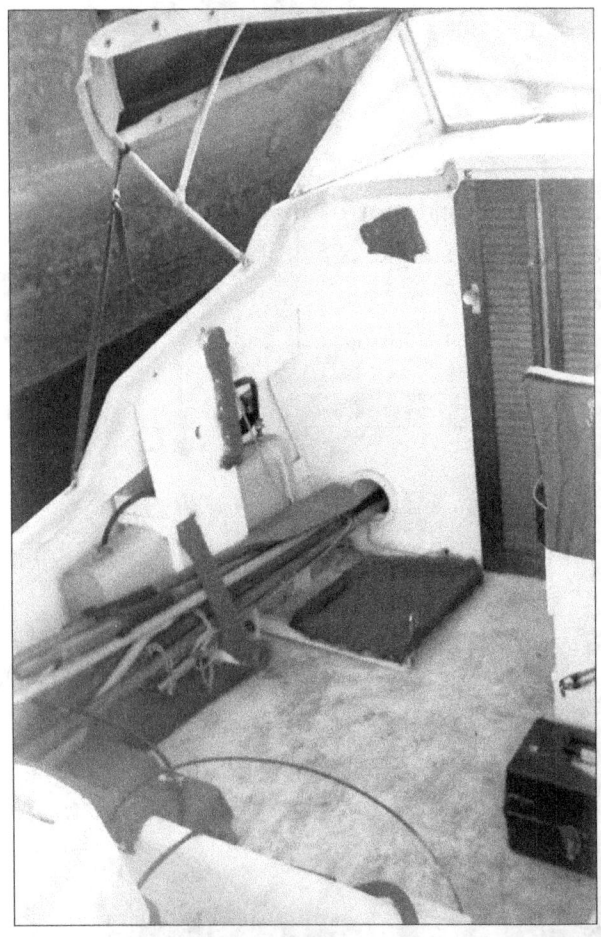 

Oar and mast storage and pads for carrying dinghy on Dotline. | Dinghy secured in cockpit on Dotline.

Dotline cockpit area in use.

Carrying dinghy with Dotline afloat.

Cockpit Size

Some stock fiberglass sailboats, especially ones that are intended for use in unprotected waters, have cockpits that are too large from a safety standpoint.

One method that I have seen used to correct this problem is to mold in a section at the forward end of the cockpit that forms a seat with a locker below that is accessed from the cabin.

This modification requires considerable fiberglassing skill, so it should only be attempted by experienced fiberglassers. The modification should be on a par with the quality and appearance of the rest of the boat.

In making this modification, a male mold can be formed from various materials, such as plywood and sheets of rigid foam. The wood or foam pieces are cut to shape and epoxy glued into position. The fiberglassing is then done over this. The example shown in the drawing can be modified to fit your particular needs.

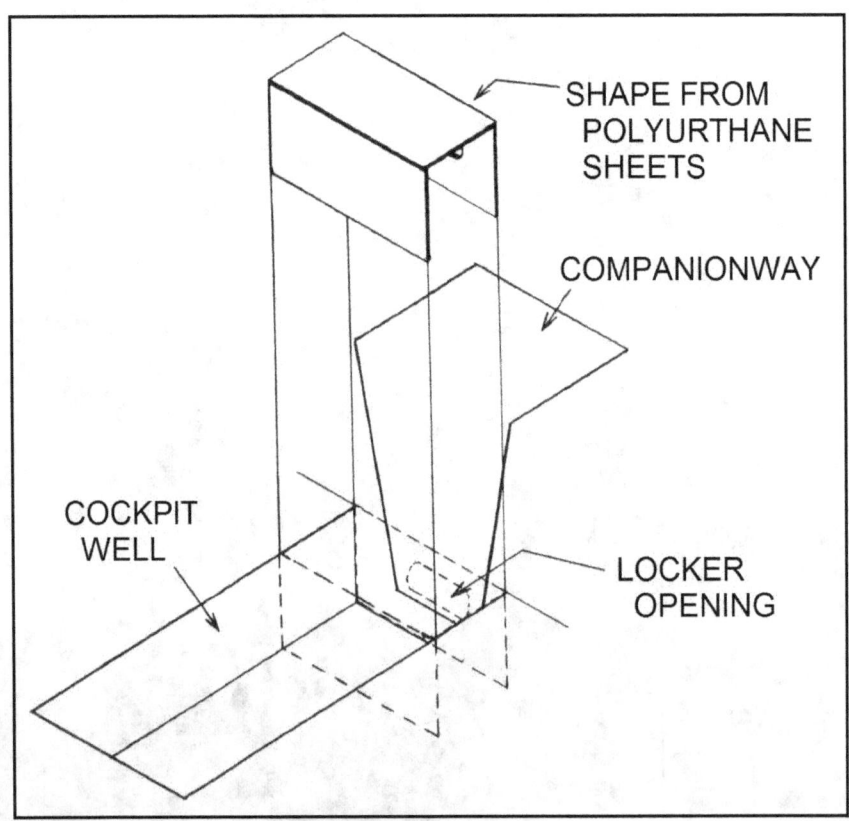

Decreasing the size of cockpit well.

Seat-Locker Lids

Many manufactured boats have cockpit seat lockers that are not watertight and/or do not have secure latching devices. Many do not have any latching device at all. At extreme angles of heel, these can come open. Adding latches is especially important if the locker drains into the bilge.

Leaking can often be corrected by adding a gasket between the rim and lid and adding a securing device such as a lever hasp or a dog clamp. For the gasket to prevent leaking when it is completely submerged, it should be held in compression.

Companionway Sill Height

In order to make it easier to get below – which in turn makes it easier to sell boats, at least to newcomers – many stock fiberglass boats have companionways that extend far too low, in some cases, to within 6 inches of the bottom of the cockpit well. For an offshore boat, it's recommended that the lowest point be above where water would begin to flow over the coamings onto the deck.

The simplest way to correct this problem is generally to seal in the lower companionway drop board with bedding compound. In some cases, a backing board is added to reinforce the drop board.

Many builders prefer to fiberglass the area in and construct a new bottom for the companionway sill at a higher level. In this process, the wood members are cut away below the new companionway height, and the area is fiberglassed in with a thickness and strength equal to or greater than the surrounding area. After finishing and matching color, wood members can be shaped and fastened in place. In some cases, the wood removed can be reused at the new location. Other modifications described in later chapters, such as those to companionway ladders, may be required in conjunction with raising the height of the sill.

Perhaps the best modification – though a difficult one – for increasing the companionway height is to add a complete fiberglass bridge between the cockpit and the companionway. The techniques are similar to those described above for decreasing the cockpit size. Remember that modifications of this scope require considerable fiberglassing skill.

6

RUDDERS AND STEERING SYSTEMS

A variety of rudder and steering systems are used on stock fiberglass boats. Traditionally, sailboats have been steered by a rudder controlled by a tiller or a wheel steering system. Powerboats are steered by a rudder and/or control of the engine thrust from propeller or water jet drive.

Typical modifications include beefing up the existing system and changing to a different system. Many owners of offshore sailboats would like to construct an emergency tiller and rudder system. Modifications such as these are the major concern of this chapter. Alterations such as changing the size and shape of the rudder, adding a skeg, and relocating the rudder involve design and construction problems that are beyond the scope of this book. Amateur builders who want to make these changes on a boat are strongly recommended to consult with a competent naval architect.

A number of vane steering systems are now used on sailboats. Frequently, these can be obtained in kit form for the owner to install. Various types of automatic pilots have been used on both powerboats and sailboats; installing them is within the capabilities of most amateur builders.

For most of the steering modifications covered in this chapter, factory-made components are recommended. However, some owners with special skills and equipment can sometimes construct their own, often making replicas of manufactured versions.

Sailboat Rudders
The methods used to strengthen rudders and mountings depend on the type of rudder and how it is constructed. Extremely

reliable are outboard rudders set in a heel casting on a keel or skeg and held to the rudderpost by two or three gudgeons.

Through-hull-mounted rudder supported by skeg.

Rudder mounting with heel casting on keel.

A recent trend, however, is toward stern and through-hull mountings with no skeg or keel attachments. Unsupported through-hull spade rudders are highly susceptible to damage, especially at the point where the rudder shaft comes through the hull. Since this is a popular type of mounting on American-made boats, and one that has caused many difficulties, we will begin with this type.

The rudder shaft generally passes through a fiberglass tube from the hull to the afterdeck or floor of the cockpit well. This tube is generally bonded in place to form a sealed hole. The fiberglass bonding in this area should be checked carefully. The shell areas at both the upper and lower ends of the shaft tube should be heavily reinforced, to the point where the rudder will break before being forced through the hull. Other methods of strengthening this vulnerable hull area are beams, frames, and bulkheads. These can be of wood, glassed in place. A watertight bulkhead placed a foot or two ahead of the shaft tube separating the hull fore and aft is another possibility for added safety. Some stock boats already have this feature.

After the shell is reinforced adequately, additional bonding layers of fiberglass can be added to the tube-hull joint. Extend the laminate at least 6 inches on the tube and 6 inches out from the tube. Several cuts are needed to shape the fiberglass mat and woven roving or cloth to the contour of the shell and tube.

Modifications such as installing a larger rudder shaft are involved and costly and are therefore, though often desirable, not covered here.

A frequent weakness in stern-, skeg-, and keel-mounted rudders is in their fittings. In many cases, it's desirable to replace these fittings with larger and stronger ones. This modification involves purchasing new fittings or having them made. In most cases, the boat must be out of the water to make the installation Remove the old fittings. If the old holes don't match the new ones, the holes should be glassed in. In some cases, new backing blocks will be needed. Drill new holes, check the fit, and then apply bedding compound and fasten down.

Fiberglass rudders can be strengthened by adding additional layers of fiberglass. However, this must be done carefully in order not to change the fairness and profile of the rudder. This modification is generally done only on cruising boats, as this modification may have a slight – probably imaginary – effect on the performance of the craft.

On some small day sailers, the problem is in the mounting and fastenings rather than in the fittings themselves. An example is gudgeons fastened to a stern with screws. To modify, remove the screws, add backing blocks, drill through holes, and fasten the gudgeons in place with through bolts. Bedding compound, of course, should be used.

Sometimes, transoms have not been adequately reinforced to withstand the stresses of the rudder, which are considerable. If this is the case, the fastening for the gudgeons should be removed and the transom reinforced on the inside of the boat. There are several ways of doing this. In some cases, additional layers of fiberglass are all that are needed. However, given the same amount of added weight, fiberglassing over a hard– or soft-core material will often result in a stronger reinforcement. Other possible modifications are adding knees, a keelson, or gunwale.

Sailboat Tillers and Other Steering Systems

Prefabricated wooden tillers are available; or one can be laminated. A clamp mold, as detailed in Chapter 2, can be used for forming a wood tiller. After laminating, the tiller is shaped, sanded, and finished as desired.

Several types of tiller extensions are now being manufactured. Installing them is generally quite easy.

Various types of wheel steering are now seen frequently on sailboats. Many owners wish to change over to wheel steering, or to have both tiller and wheel steering capabilities. Three basic systems of remote-from-the-rudder wheel steering are used: drum, push-pull, and hydraulic. The choice is up to the individual, although price will probably be a consideration; hydraulic systems are expensive.

Two other wheel systems, which must be located near the rudder-head, are worm gear and geared quadrant. If possible, try a similar boat equipped with the steering system you are considering before adding it to your boat. Some of the systems, such as hydraulic, give almost no feel.

A simple drum system with cables that can be attached to the tiller for wheel steering in stormy weather in a doghouse or under a dodger has been used successfully on many cruising sailboats.

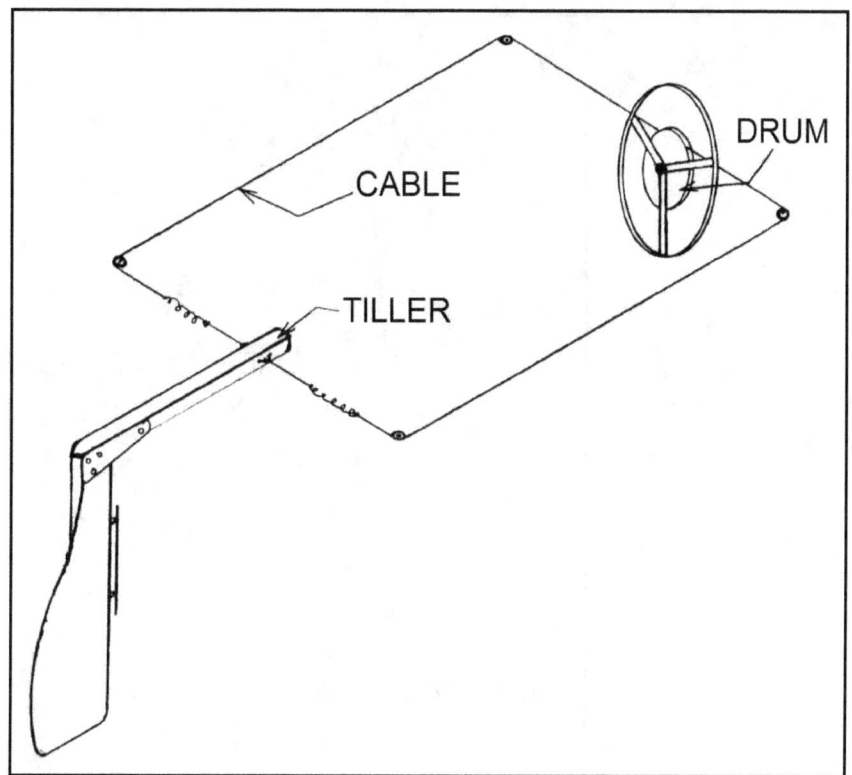

Drum steering system that is attached to tiller.

Pedestal steering is popular among some boat owners. Complete kits for installing this type of steering are now available. The steering wheels are usually of the type known as destroyer wheels. These can be covered with leather, such as elk hide, so that they will not be cold and slippery.

Emergency Steering

In sailboats used offshore, there should be some method of fitting an emergency tiller. This is especially important in the case of through-hull mountings, which cannot be readily repaired at sea. One way of making an emergency rudder system is to fabricate a separate rudder and tiller that will fit gudgeons mounted on the transom. This way, the emergency rudder can be attached at sea if the primary system fails.

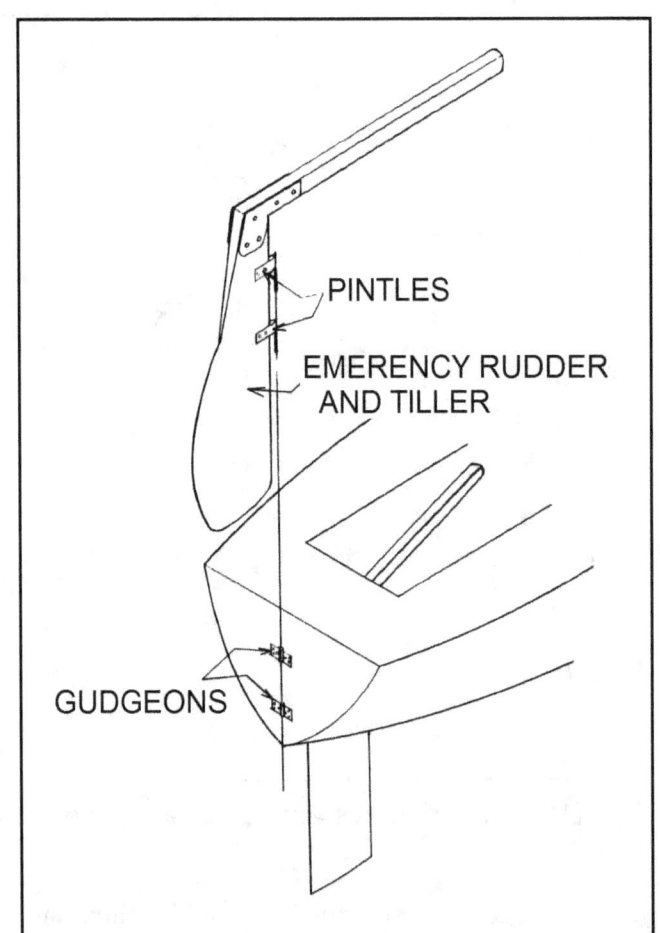

Emergency rudder for sailboat.

Likewise, in sailboats with wheel-steering systems there should be a method for mounting an emergency tiller in the event that the wheel system fails.

Self-Steering Systems for Sailboats

Many self-steering vane systems have been used on sailboats. A number of proven designs are now manufactured. These come with mounting instructions, and installation is generally easy and straightforward.

The main problem is in selecting the self-steering system that is right for your boat. The prices range from a low of about $125 up to a thousand dollars or more. The low-cost units generally have lines running to the tiller for making course corrections according to the wind direction. The more expensive ones have a trim tab that connects to the rudder or to a separate rudder. The low-cost units can be highly satisfactory on some boats. They have been used successfully on a number of long ocean passages. However, they are unsatisfactory on other boats.

Try to find out what types have proved satisfactory on the model of boat you own. Often, the manufacturers have information about this.

A number of automatic pilots for wind, compass, and GPS courses are now available for simple installation in sailboats. Many of these claim low power drain.

Powerboat Steering Systems

Except for some small outboards, powerboats generally have wheel steering, which is accomplished by controlling the rudder or the thrust from the engine. Most rudder systems are used in connection with the engine thrust – that is, the propeller is forward of the rudder and the rudder thus controls the direction of the thrust, at least to some extent. The steering of outboards and inboard-outboards, as well as jet-drive units, is generally accomplished by controlling the direction of the thrust alone.

Whether the direction of the boat is controlled by the rudder and/or by the thrust from a propulsion unit, three basic types of linkage to the wheel are used: drum, push-pull, and hydraulic. Many boat owners want to change from one system to another. Each has advantages and disadvantages. All of them, perhaps excepting the drum system, are best purchased as complete units in kit form.

The simplest and least expensive system is the drum. The steering wheel is backed up with a drum, which has cables wound around it leading through sheaves along one or both sides of the boat to the outboard motor or rudder quadrant. Although the drum system is subject to wear and, thus, breakdown, it's easy to repair.

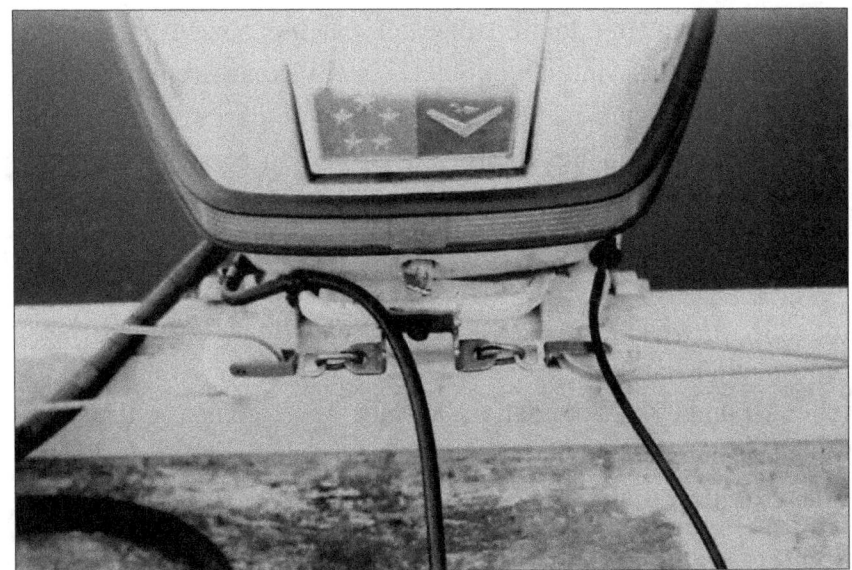

Steering cables and pulleys connected to outboard.

Push-pull steering is popular among boat owners with high horsepower outboards and stern-drive units. Movement from the steering wheel is transmitted to the rudder by a stiff wire within a flexible housing. The system can be used for dual-control stations. This method can also be used for controlling dual-engine stern drives with a single steering wheel, provided the horsepower is not too great. In some cases, the two units can be linked in tandem to a single push-pull cable.

The most costly systems are hydraulic. These have been used on larger boats for a number of years, and units for smaller boats are now readily available, for a price.

These, then, are the basic choices. By following the instructions that come with the units, you should have no special problems with installation. It is recommended that components be mounted firmly with through bolts. In most cases, backing blocks should also be used.

Steering wheels can be located as desired, but they should be mounted firmly. Many types and sizes of steering wheels are available, including spoked character wheels and destroyer wheels. They are made from a variety of materials, including mahogany, teak, stainless steel, bronze, aluminum, and various plastics.

7

ENGINES AND MOUNTINGS

Typical modifications include adding brackets or wells for outboard motors and changing from outboard to inboard or from one inboard to another. In all cases, remember that the power requirements for a particular boat are inherent in the design. The boat generally includes a range of options, and a good guide to power requirements is what the manufacturer of the particular boat offers in the way of standard and optional engines and mountings. Besides regular outboard and inboard engines, there are inboard/outboard (I/O) units, including outboard-type motors with through-hull drive units so that the motor can be mounted inboard.

Mountings for Outboard Motors

The basic motor-mounting methods are: directly to the transom, often in a transom cutout; on special mounting brackets, which may be sliding, folding, or detachable; and in outboard wells.

Depending on the length of shaft on the outboard, a transom cutout may be necessary. In order not to reduce the freeboard, a false transom or well can be added on some boats. The installation of a false transom is similar to that of a bulkhead.

A well in the shape of a self-draining box can be constructed of plywood and covered with fiberglass. Drain holes through the transom, which should be lined with fiberglass or through-hull fittings, complete the modification.

Any cutouts made in a fiberglass transom should have fiberglass strips added over the exposed ends of the laminate. Protective pads of wood, metal, or plastic can be used to keep the motor clamps from damaging the fiberglass.

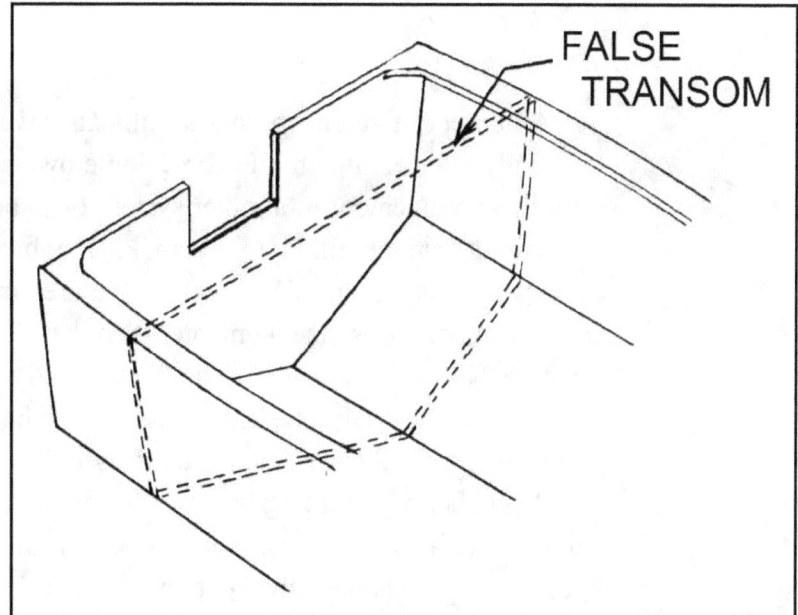

Mounting a false transom.

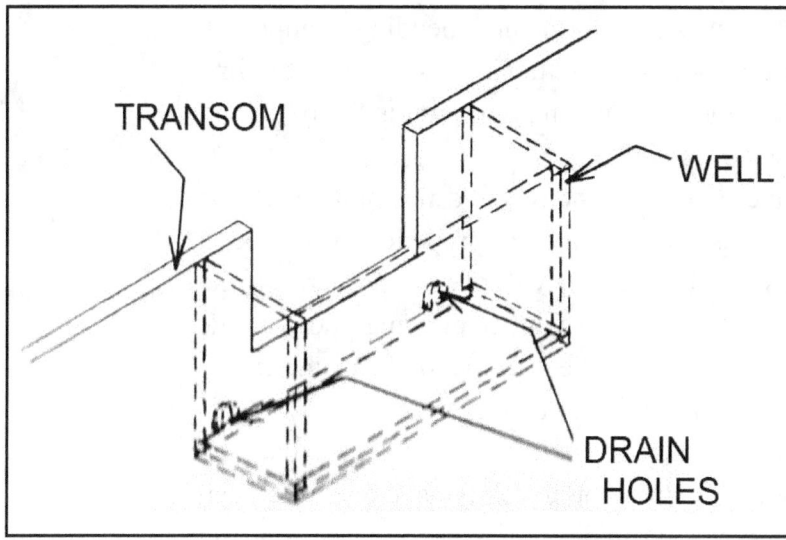

Box transom motor well.

Box transom motor well that extends all the way across the boat.

A number of stock fiberglass sailboats have a deep cutout in the transom for taking an outboard. Some owners want to change from this arrangement to another system or to an inboard engine. The cutout can be filled in by making a backing mold from cardboard or wood and fiberglassing the area in to the adjoining shell. In most cases, the laminate should be approximately the same as that of the transom. Another method is to shape a rigid foam block and then fiberglass over this. The foam should be polyurethane if polyester resin is to be used.

A number of mounting brackets for outboard motors are now being manufactured. These may be solid-mounted, on sliding tracks, or detachable. Installation involves strengthening the transom and then through-bolting the bracket or mounting plates in place, using large backing blocks and bedding compound. When installing the type with parallel tracks, it's important to line up the tracks exactly; otherwise, the motor mount is likely to bind when the unit is raised or lowered.

The height of the mounts in the lowered position should be considered carefully, as most outboard propellers must be a certain depth in the water before they will operate properly. If the propeller comes out of the water with a pitching motion, the engine is likely to wind out and die. A long shaft on the motor will allow higher placement of the mounting bracket.

Several modifications can be made on motor wells; or a well can be added to a boat that does not presently have one. The open hole at the bottom of the well can decrease performance. To avoid this, some boat owners use a plug that fits in place when the engine is not being used.

Sealing in a motor well is a modification that is sometimes done. Generally, if a change is made to another method of mounting an outboard, or if inboard power is added, the well should be sealed in. This might involve only bedding the present plug in place. Other owners fiberglass over the opening, matching it with the hull shell. If the well itself is removed from the boat, the patch over the hole should be at least as strong as the adjoining shell.

Adding a motor well can present tricky design problems on many boats. It's a good idea to check with the manufacturer or a designer to see if what you have in mind will be satisfactory. In some cases, it will be necessary to locate the well off to one side of the center line. Wells are typically located inside a stern locker or through the cockpit floor. Full bulkheads fore and aft of the well are typically used. An outboard operated in a locker has the same ventilation requirements as an inboard installation.

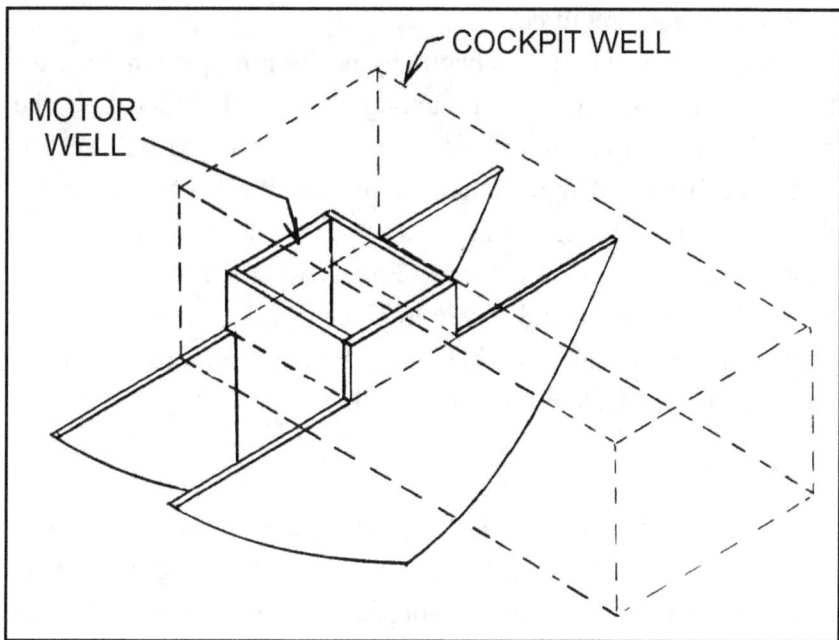

Installing a motor well through cockpit sole.

Inboard/Outboard Installations

Inboard/outboard (I/O) engines are best purchased as a complete unit, with mounting hardware. Since these units are unsuitable for many boats, check with the manufacturer to see if they can be used on your boat.

If the inboard/outboard engine is to be mounted to the transom without further support, considerable reinforcing and strengthening of the transom is generally necessary. Many units also have an engine bed, similar to those used on regular inboard installations. Methods for adding engine beds are covered later in this chapter.

A typical inboard/outboard package consists of the complete engine; I/O drive unit; instruments and panel; steering, control, and ventilation systems; tanks; mounting hardware; necessary templates; and complete installation instructions. In most cases, the installation is fairly straightforward. Two or three persons can generally handle a small unit without a hoist. For larger units, a hoist is generally required.

A typical installation is begun by positioning the template for the transom opening. The mounting holes are drilled and the opening for the drive unit is cut. A saber saw can be used for this. The I/O drive-unit flange is then mounted in the transom opening. Bedding compound should be used for this. The drive unit is then positioned and bolted in place. Following this, the engine is lowered into position and bolted in place. The ventilation system, instrument panel and wiring, steering and control systems, battery, and fuel tank are then installed.

Inboard Installations

Our concern here is mainly with the modifications to the boat that are needed if you install an inboard. Most engine makers provide detailed installation manuals; these should be followed closely.

Usually, the manufacturer will give you information about suitable inboard power units. Weight, horsepower, size, and shape are important considerations. Often, the manufacturer can provide plans showing the exact location of the center line of the propeller shaft, placement of the engine bed, position of mounting bolts, and so on.

Some stock fiberglass boats have engine beds even though no inboard was previously mounted. If this is not the case, metal engine bearers can be bolted to wooden engine beds. To construct a wooden engine bed, first shape the wood parts. To avoid serious had spots, the area of wood-to-hull contact should be fairly large.

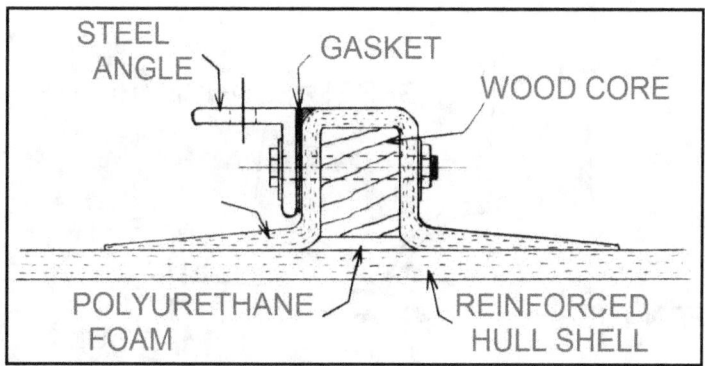

Engine bearer.

The wood members should be of a dry, hard wood. The members should end at a bulkhead or stringer.

Some builders use a soft-core material between the wood and shell contact areas; others bond the wood directly to the shell with epoxy. In either case, the hull should be well thickened in the area of installation. It may also be possible to join the engine bed to stringers or other structural members that are already in the hull. Another possibility is to connect the bed to the keel or to the top of the ballast.

The wood members are then fiberglass-bonded to the shell. Many builders completely embed the wood within a sheath of fiberglass. The engine stringers should be lined up carefully before the bonding is done.

Stern tubes are mounted in various ways. A typical installation is shown. In this installation the tube is fitted in holes, which must be carefully positioned, in the hull shell and bulkhead or frame. The tube is wrapped with fiberglass and then bonded in place at each end. A thick laminate of mat and cloth or woven roving should be used. In some hulls the space above the shaft tube can be filled in with resin mixed with micro-balloons. Make sure, however, that there will be easy access to any fastenings that are to be used. The filled-in area can then be covered over with fiberglass.

If a propeller-shaft bracket is to be installed, the hull should be heavily reinforced. This is generally done inside the hull shell. A large backing block should be used for through-hull fasteners.

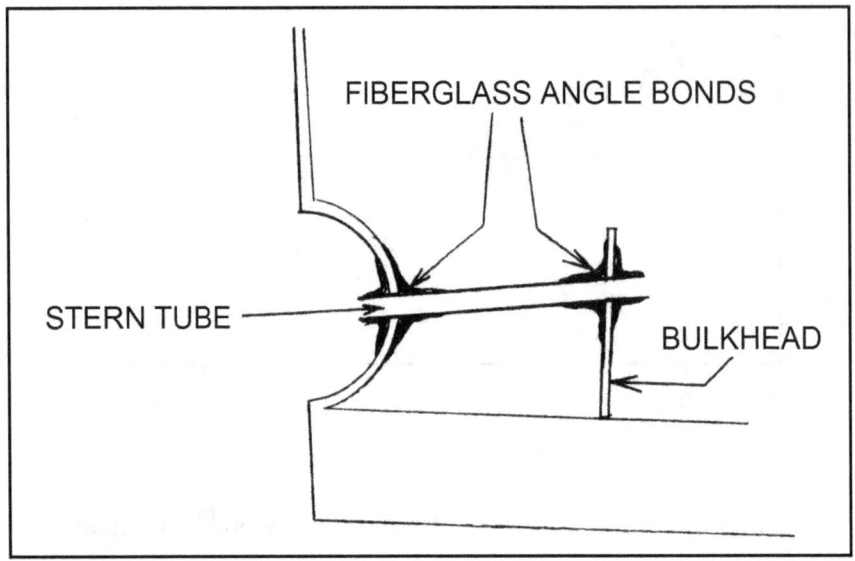

Stern tube wrapped with fiberglass and bonded in place.

Installation of components, such as the shaft coupling, propeller shaft, and stuffing box, varies with the engine used and the type of boat and engine mounting. The manufacturer's instructions should be followed carefully.

A hoist of some kind is needed for all except very small engines, which can be lifted into place by two or three people. Hoists are generally available in boatyards. In some cases, a jury-rig can be constructed.

The type of engine mounts and the method of fastening them to the engine stingers vary. Flexible mounts used in conjunction with a flexible engine coupling commonly allow for vertical adjustment, thus eliminating the need for shims and wedges when lining up the engine and propeller shaft.

Follow the manufacturer's directions when installing fuel, exhaust, electrical, control, and ventilation systems. Several through-hull fittings are necessary. These generally require reinforcing the hull and using backing blocks for installing them.

8

EXTERIOR FITTINGS AND TRIM

The exterior fittings and trim on many stock fiberglass boats leave much to be desired. In many cases, however, modifications can be made to improve on these shortcomings. Typical desired changes include adding a fitting where none is attached and replacing an existing fitting with one that is stronger or of a different type. Adding trim pieces can improve the appearance of a boat and serve valuable protective functions. Stanchions, lifelines, and pulpits can add to the safety of the craft, and a canvas dodger makes the helmsman's job more comfortable in foul weather. These exterior modifications, as well as many others, are the subject of this chapter.

Cleats

Attaching a cleat to the shell involves reinforcing the shell with additional layers of fiberglass, shaping and embedding a backing block, and drilling holes for the fastenings. After the cleat has been dry-fitted, bedding compound is applied, and the

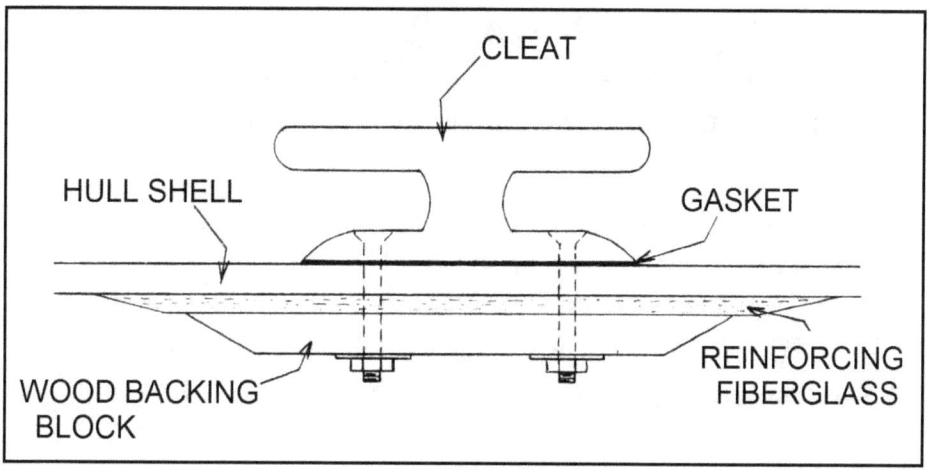

Mounting cleat to hull shell.

cleat is bolted down. For heavy cleats and those with small areas of contact with the shell, a thin gasket of neoprene or rubber should be used as a cushion between the cleat and shell to minimize compression damage. Large washers should be used on a wooden backing block. Care should be taken not to over tighten the nuts.

To add a cleat to a sandwich shell with a soft core, a solid fill or plugs must first be bonded in place. If this is not done, a load on the cleat will cause compression of the laminate, and the cleat will become loose, with leaking likely to result.

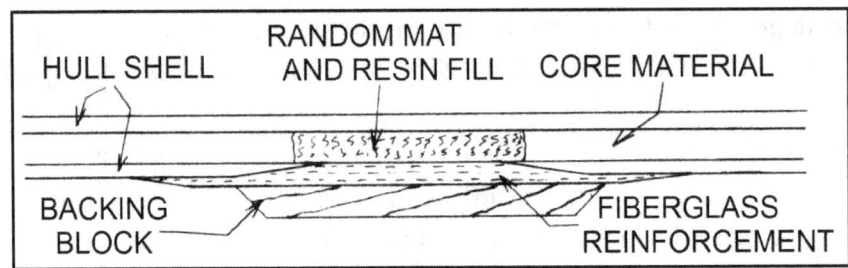

Solid fill for attaching fitting to sandwich-core hull shell.

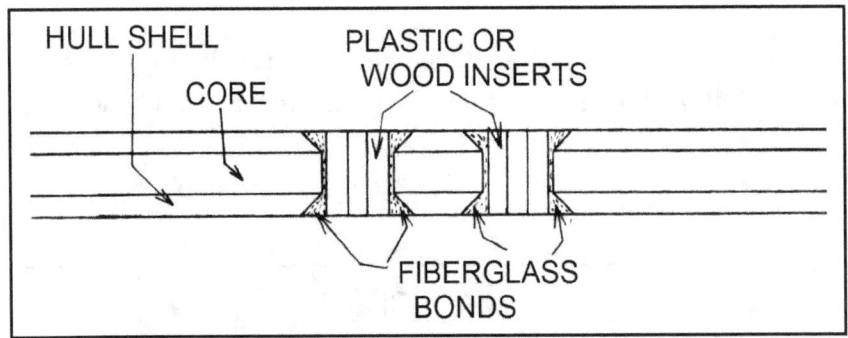

Plugs for attaching fitting to sandwich-core hull shell.

The most important rule to follow in installing cleats is that any strain, regardless of cause, should allow the fitting or its fastenings to fail before the boat.

To replace a cleat with one with different spacing for the fastenings, it may be necessary to fiberglass in the old holes. By fitting the backing block first, filling in the old holes will be easier, as a convenient backup is formed. Larger holes, however, should have reinforcing layers of fiberglass placed behind the patch, before the block is embedded in a single layer of wet mat against the shell.

Stanchions and Lifelines

Stanchions should be mounted in such a way that they will bend before tearing away from the deck. This requires reinforcing the shell with additional layers of fiberglass and large backing blocks.

Stanchions sometimes have separate bases. Components for lifeline assemblies can be purchased separately or as a complete kit. Stanchions come in various lengths for both single and double lifelines. If longer stanchions than those presently on the boat are desired, new ones can often be purchased to fit the old bases; this also applies to converting from single to double lifelines, although an easier and cheaper solution – if the present lifelines are long enough – is to drill holes for the lower lifeline in the present stanchions. Always be sure to use through bolts when fastening down the bases.

Lifelines are generally attached forward to a bow pulpit. If the boat has a stern rail, the lifelines are generally attached to it; if there is no stern rail, the lines are generally fastened to deck tangs. Pelican-hook gates are often included in the lifelines for easier boarding.

Bow Pulpits and Stern Rails

Rails can be purchased in many standard measurements, or they can be made to special patterns by some companies. In most cases, standard measurements can be used. In selecting rails, note that various materials and thicknesses are available. For most installations it's advisable to use stainless-steel rails rather than the lower-priced aluminum ones, especially if the boat is to be used in unprotected waters. Safety devices should be constructed so that they will not fail when they are most needed.

Mounting rails is similar to mounting stanchions. The fiberglass deck should be reinforced, and backing blocks should be used. Although rails do not give way as frequently as stanchions, they are subject to considerable stresses.

Boarding Ladders and Steps

Several types and designs of boarding ladders are commonly used. Some of these merely hook over a railing and require no installation. Others have mounting brackets that are attached to the stern, hull, or deck; these should be through-bolted, and backing blocks should be used. In some cases, the fiberglass shell will have to be reinforced with additional layers of fiberglass. In most cases, the ladders are removable for easy storage.

Many sizes and shapes of boarding steps, both stationary and folding are available. Attached low to the stern of a boat, these steps provide an emergency means of getting aboard. Round handles also serve this function. These fittings should be through-bolted with backing blocks. Bedding compound should also be used.

Grab Rails

Grab rails can be purchased in various styles and lengths at marine supply houses. Teak is generally the preferred wood. Grab rails should have suitable handhold cutouts. Solid grab rails should not be used.

Some builders shape their own grab rails. Two grab rails can be shaped from a single piece of wood. Teak is the preferred material. The ends are rounded with a band saw or saber saw. Three-inch holes are drilled and the sections between each pair of holes are cut out. A router is used to round all corners, both

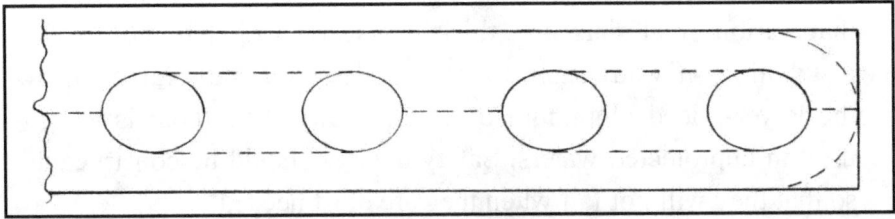

Construction of a pair of matching grab rails from a 1-by-6 board.

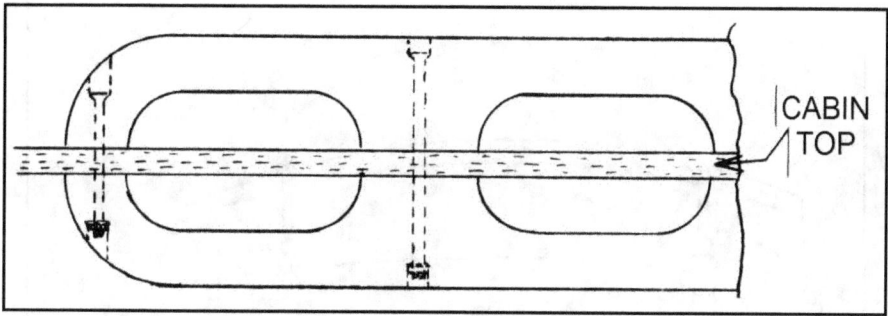

Mounting interior and exterior grab rails opposite each other.

around the board and in the cutouts. This is done on both sides of the board. Cut the board into two matching grab rails.

Preferably, all fastenings should be through bolts. On the rail, the bolt heads should be countersunk and capped with wooden plugs. On the interior side, large washers with cap nuts can be used where appearance is important, but a wooden or metal backing strip is generally a better arrangement. Another possibility is to position the interior and exterior grab rails opposite each other. This conveniently solves the problem of interior appearance.

Dinghy Cradles

These can be shaped to carry the dinghy upright or upside down. The upper edges of the cradle should be padded to protect the dinghy. The cradle supports should be through-bolted with backing blocks and well bedded with a sealing compound.

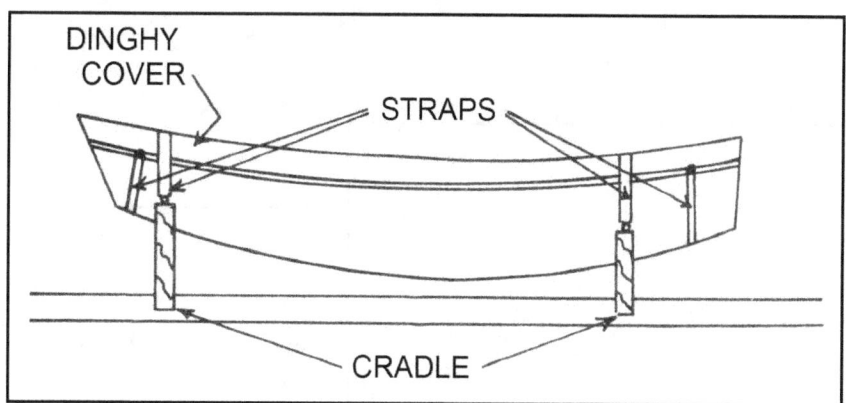

Cradle for carrying dinghy upright.

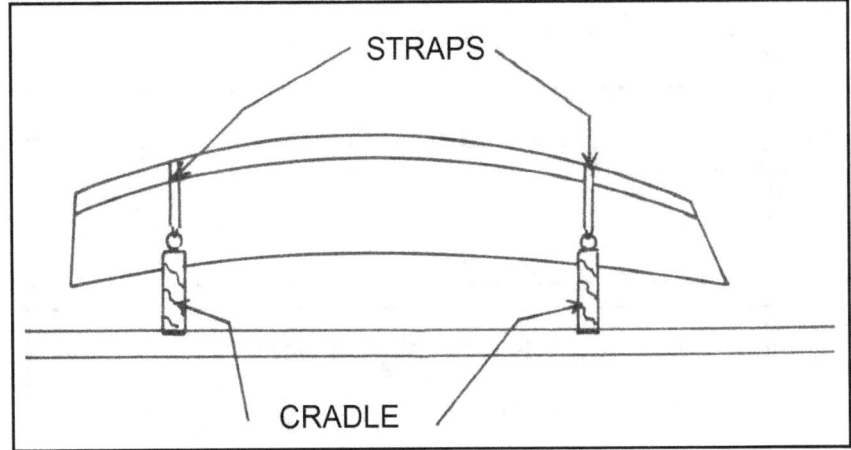

Cradle for carrying dinghy upside down.

Two ways to mount a dinghy over a companionway hatch are illustrated: a bridge and a mounting to a hatch hood. Both methods allow the hatch to be opened with the dinghy on the cradle.

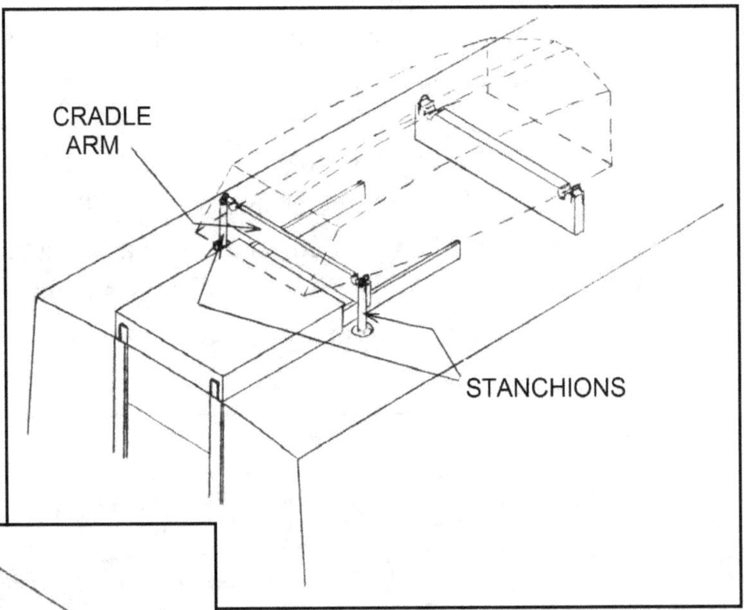

Cradle arm attached to stanchions to form bridge over sliding hatch.

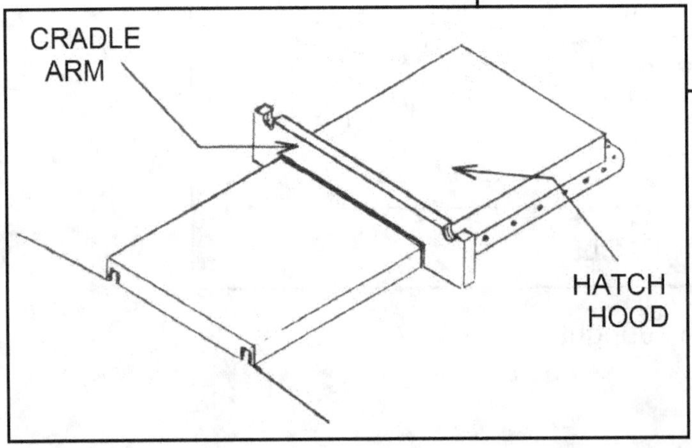

Cradle arm attached to hatch hood.

Dodgers and Awnings

Manufactured fittings are readily available. The dodgers and awnings can be made from heavy canvas or Dacron. You can have these made up at a shop, or you can make them yourself if you have the necessary sewing skills and equipment. The sewing can be done by hand, but for large jobs, this can be time consuming, so machine sewing is generally done. Grommets and fasteners are available at marine supply stores. If fasteners are to be attached to the boat shell, the kinds with plates for through bolts are recommended.

Dodger that I made for a Dreadnought 32.

Awning that I made for GO FOR IT.

Windshields

The primary purpose of most windshields used on boats is to deflect spray. Thus, they are sometimes called "spray shields." Windshields are popular on runabouts, speedboats, and other powerboats, and are sometimes used on sailboats.

Though windshields can be home-fabricated, a manufactured version is recommended, as the money saved by making your own is likely to be small. In addition, giving it a neat appearance can be difficult. As a minimum, I suggest that the brackets be purchased. However, if a wraparound type is desired, it is best purchased as a complete unit. Styles and sizes that fit most stock boats are readily available.

Windshield brackets should, in most cases, be through-bolted and backing blocks should be used. Bedding compound should be applied between the shell and the brackets before the final tightening down.

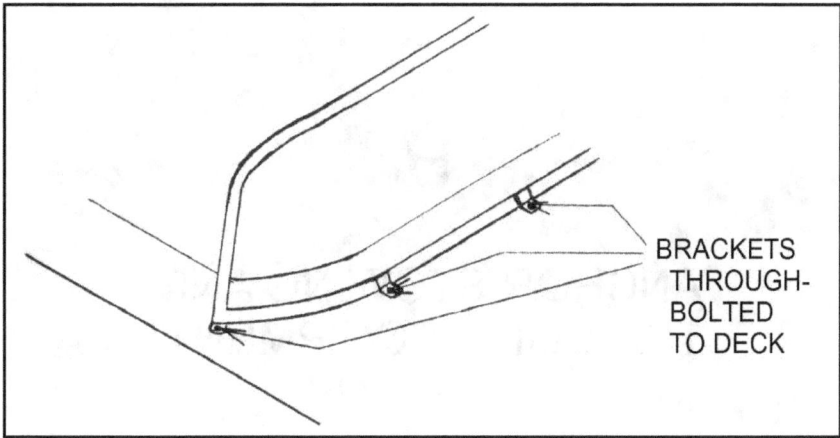

Installation of manufactured windshield.

Folding Boat Tops

Construction and mounting is similar to that of dodgers, except the top attaches to a windshield. Complete kits for folding boat tops for many stock boats are available.

Folding boat top.

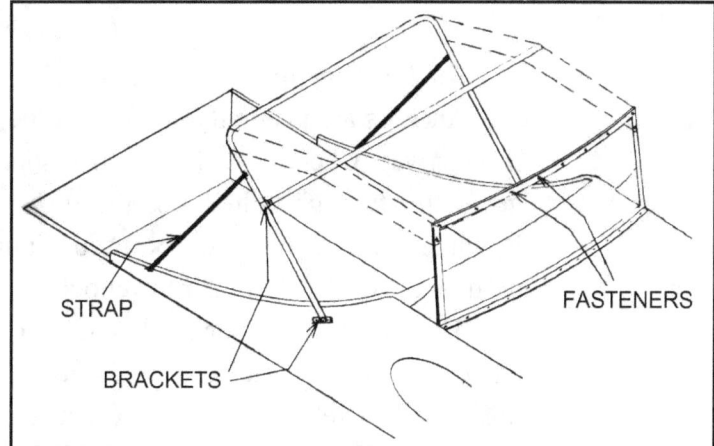

Windshield and folding boat top I fabricated and installed on cabin top of Dotline.

9

ANCHOR STOWING AND HANDLING EQUIPMENT

On many stock fiberglass boats, all anchor stowing and handling equipment beyond a cleat and fairleads are optional extras. The result is that many boats afloat have no means of stowing anchors and rodes so that they can be broken out and put over quickly. Light anchors can be handled by hand. Heavier ground tackle calls for mechanical aids, such as windlasses, capstans, and davits. Equipment that can be operated manually or automatically is available.

Anchor Stowing Equipment

Anchors are generally stowed on the foredeck when the boat is underway. Chocks that fit most anchors are available. These are generally through-bolted to the deck with backing blocks. Bedding compound should be used. Sometimes, blocks are also used on the deck side under the chocks.

The location of chocks should be considered carefully. If a chain and/or rope deck pipe is used, the chocks should be positioned in relation to this so that a minimum of chain is on deck between the deck pipe and anchor. If this is not feasible, a means to protect the deck between pipe and anchor should be provided. One method is to install a sacrificial strip of teak or a wooden trough. A metal channel is another possibility; manufactured ones are available.

A number of bow chocks and anchor rollers are designed for anchor storage. These can be made at machine shops if no suitable manufactured model is available. Many stock fiberglass boats manufactured in England come with bow rollers that are designed for holding plow anchors, such as the CQR.

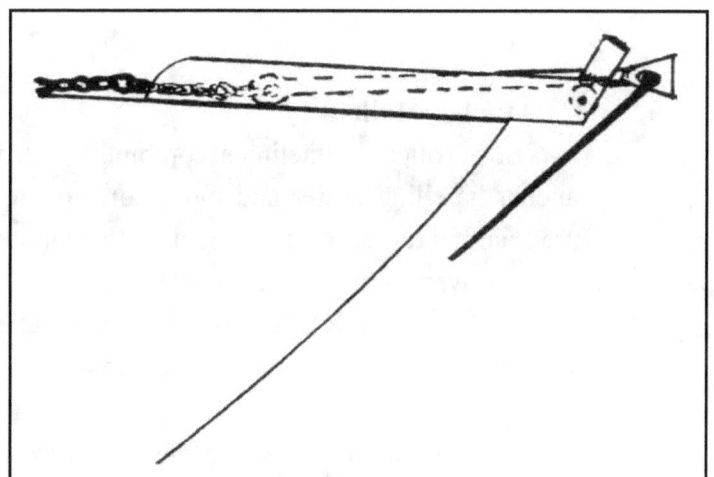

Anchor roller that allows anchor stowage.

Boats with bowsprits often have anchor chocks or rollers mounted on the bowsprit. This often provides a convenient way to stow anchors. Sometimes the anchor is swung aft from the chock and secured in a special clamp. Bowsprits are sometimes mounted on boats especially for the purpose of mounting an anchor roller and stowing the anchor.

Anchors are sometimes secured to a bow pulpit. Anchors are sometimes secured to the pulpit with lashings. Special clamps for Danforth anchors are now available. These come with mounting clamps and are simple to install.

Special clamps for stowing anchor.

Anchor Rollers

Bow rollers, sometimes combined with a chock, can make anchor handling easier and more convenient. A variety of these are manufactured. They are generally mounted so that the roller extends over the bow. A type that folds back on deck when not in use is also available. Bow rollers are mounted by through bolts with a backing block. Bedding compound should, of course, be used.

Rollers can also be installed for stern anchors. This is convenient for letting out a second anchor or as an emergency anchor system.

Chain Stoppers

If a windlass or capstan is not used, ratchet devices can be used for stopping and holding chain. In catalogs, these are described variously as stem-head, automatic, and eccentric-cam ratchets. Mounting is generally by through-bolting a plate to the deck. Bedding compound and a backing block should be used.

Manual Windlasses and Capstans

A capstan is generally mounted vertically; a windlass, horizontally. However, the terms are sometimes used interchangeably. A capstan is generally operated by a crank; the windlass, a handle. A windlass requires the use of an anchor chock and may be single-action (takes up chain only on pulling stroke) or double-action (takes up chain on both pulling and pushing strokes). A number of models are on the market. Most take one size of chain (generally 1/4-inch, 5/16-inch, or 3/8-inch) and have a drum for the rope. A typical mechanical advantage is 12 to 1.

Manual windlasses and capstans are ideal for boat 25 to 35 feet LWL, especially if the boat is to be used for cruising. They have also been used on both smaller and larger boats.

When mounting windlasses and capstans, the deck must be well reinforced. In addition, the fastenings should connect to a beam or bulkhead, directly or indirectly. A large backing block should be used to spread the load over a large area.

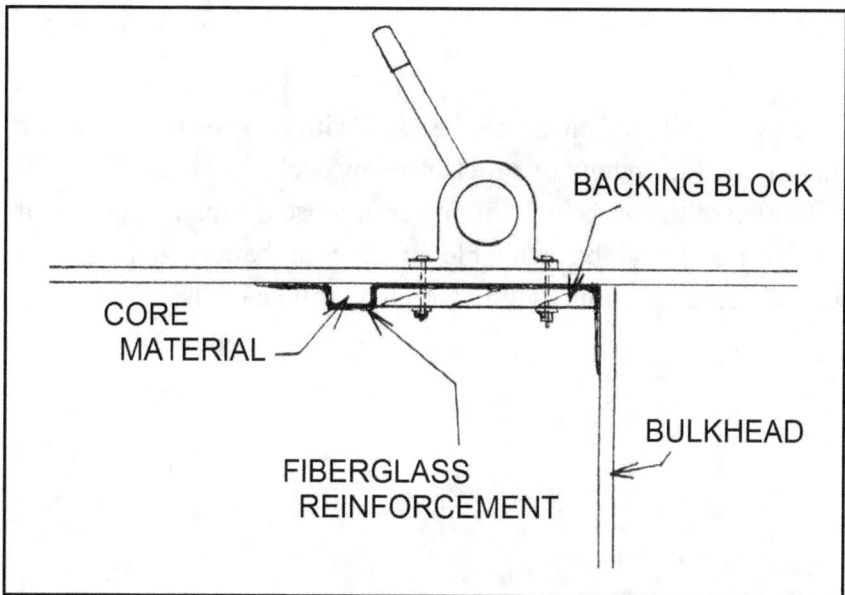

Mounting windlass between beam and bulkhead.

Windlasses and capstans are generally located close to the chain locker. If this is not the case, sacrificial strips of wood or a metal channel should lead to the locker to prevent chafing of the deck.

Power Windlasses and Capstans

These are available in electric and hydraulic models. The pump for the hydraulic system is generally belted off the engine, and it pumps hydraulic fluid through pipes and/or hoses to power the windlass or capstan. Hydraulic units often have an efficient hand-power backup arrangement; electric models seldom do.

Some power units are mounted and used like manual ones. Others are part of automatic anchoring systems that can be operated from a remote-control panel.

Electric and hydraulic windlasses and capstans are best purchased as a unit with the necessary mounting hardware. Electric models that operate on various voltages and that are generally easy to wire are available. Mounting is similar to that of the manual units described above. Also available are hydraulic units with pumps that are easy to attach to most engines. The automatic units typically have a system for stowing the anchor.

The nylon rode (chain cannot be used with most of the automatic units) is led in and out of a locker below deck.

Regardless of the type of power unit used, a manual means of anchoring should be available in case of battery failure with electric units or engine failure with hydraulic models.

10

HATCHES AND VENTILATORS

A common shortcoming on stock fiberglass boats is hatches that leak and/or are not strong enough. Most hatches are molded from fiberglass. Sometimes they are of wood. A recent trend is to construct opening hatches from metals, such as aluminum and stainless steel. There are many modifications that can be made to improve hatches, from simple installations of gaskets to the construction of complete new units.

Ventilation is often neglected on stock boats. At best, it's generally an optional extra. Yet, good ventilation is an extremely important part of any well-found boat. The purpose of ventilators is to let air in while keeping water out. Many designs and types of ventilators have been used. Common modifications include replacing a ventilator with a better one and adding ventilators.

Opening Hatches

We will first consider hinged and clamp-on hatches. Most hinged and clamp-on hatches on stock fiberglass boats are of molded fiberglass. There is little possibility of them leaking through the fiberglass, and most of them effectively keep rain water out when they are closed. However, many of them leak when submerged in green water.

Another problem with hatches is that many of them are not really strong enough. They should be able to withstand the weight of a man jumping on them. Many hatches fail this test. Other weaknesses of hatches are in their hinges and catches. Also, they should have a strong clamping or securing device if the boat is to be used offshore.

The first modification to be considered is alterations to the present hatch. In some cases, a leaking problem can be corrected by adding a gasket and/or improving the securing device. To be effective, the gasket must be in compression all around. A vee-shaped edge forced into the gasket is generally a good way to do this.

If altering the existing hatch does not solve the problem, consider modifying the coaming. For example, a single coaming can be changed to a double one with drain holes on the outside channel.

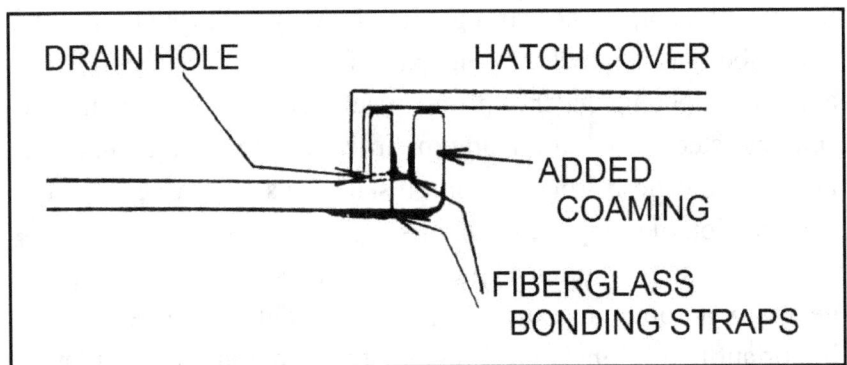

Adding second coaming to hatch.

There are a number of ways to strengthen a hatch. A fairly easy modification is to add additional layers of fiberglass to the inside of the hatch molding. To do this, first remove the hatch. The layers must be added so as not to effect the seating arrangement. If a hatch molding is extremely weak, fiberglass reinforcement can also be added to the outside, but this presents difficult finishing problems, and thus is best avoided.

Adding beams or a solid section of wood to the inside or outside of a hatch is another possibility. Adding thin teak strips to the outside of a hatch can add to its appearance as well as to its strength.

Molding in fiberglass stiffeners adds to the strength of the hatch. These can be placed on the inside and/or outside. Make sure, however, that they do not affect the opening and closing of the hatch.

If the hinges are inadequate, they can be replaced with better ones. If there is no securing device, this should be added. A removable securing device for use in heavy weather is shown below. It is easy to fabricate and will hold the hatch firmly in place.

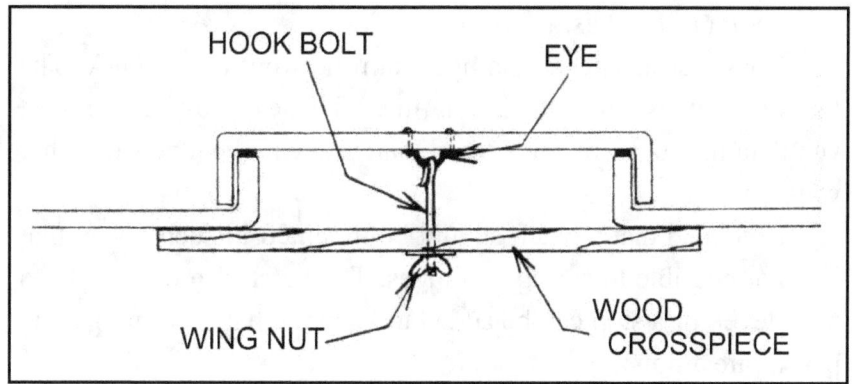

Hatch securing device for use in heavy weather.

Hatches that are clamped on from the outside, such as those used on stern deck lockers, generally provide a satisfactory seal if the gasket is in good condition and properly located, and if the securing latches hold the hatch firmly in place. Hatches that unclamp completely from the boat should have a safety line permanently fixed to them so that they will not be lost overboard.

If the present opening hatch cannot be modified adequately, a new hath can be added. An ideal way to do this is to purchase a complete hatch assembly. These are available in plastic, aluminum, stainless steel, and other metals. Some of these are translucent to allow light below. A typical size is a 20-by-20 inch opening, which allows the hatch to serve also as an escape hatch. Hatch assemblies are easy to install and come with complete instructions. Interestingly, a number of stock boat manufacturers are now using these units. Hatch assemblies are priced from about $75 to $300. This may seem expensive, but they can be easier to install, and a good job is essentially guaranteed.

Hatches can also be fabricated. Since a hatch is a simple shape, it makes an ideal first molding project. This can be done with a male or female mold. Wood construction is another possibility.

A variety of hatch hardware is available from marine stores. Do not skimp here, as hinges and securing devices are often the weak point of hatches.

Note that hatches can hinge from the front or the back. The forward hinges are safest, but the aft ones provide the best ventilation. Some manufactured hatches can be opened either way.

Note also that a hatch should be mounted in such a way that it is not possible to spring its hinges. The hatch can open right to deck level, or a stop can be added to keep the hatch from opening past a safe amount.

Sliding Hatches

Companionways typically have sliding hatches. Molded fiberglass hatches are most common on fiberglass boats. Two common hatch-slide arrangements are shown.

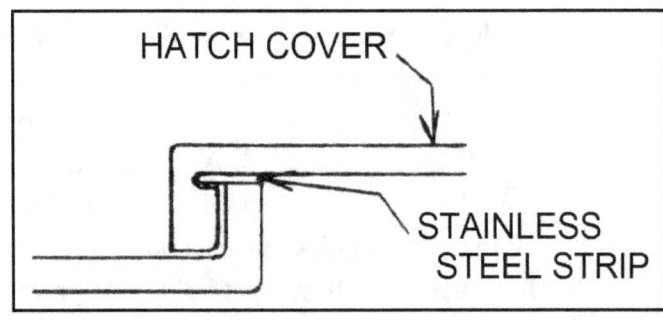

Hatch-slide arrangement with metal slide at top.

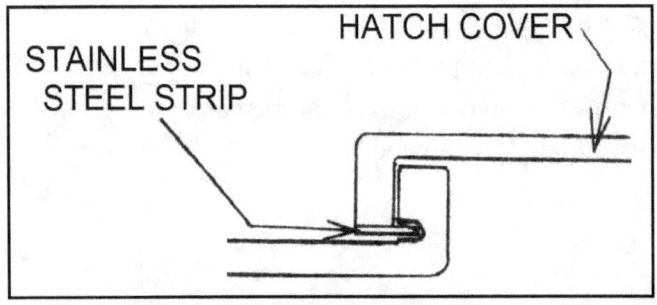

Hatch-slide arrangement with metal slide at bottom.

Hatch-slide arrangement I installed on Dotline.

Hatch Hoods

An effective way to prevent water from getting through the forward end of the hatch is to add a hood. Prefabricated ones that fit most stock boats are available, or you can fabricate your own from wood and/or fiberglass. A typical construction is shown. Allow ample clearance between the hatch and the hood, as grit trapped in a close fit will scratch the hatch. The hood should have drain holes.

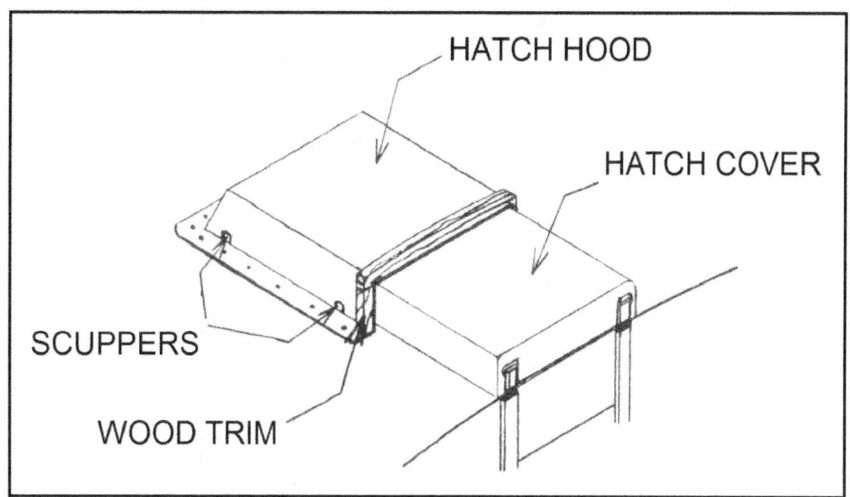

Hatch-hood installation.

Companionway Drop Boards and Doors

The drop boards on many stock boats are inadequate, especially if they are used offshore. They should be comparable in strength to the surrounding shell. Adding wood strips, as shown, is a simple way to reinforce drop boards. This modification also helps prevent water from entering. In many cases, the jambs and sill also need to be beefed up.

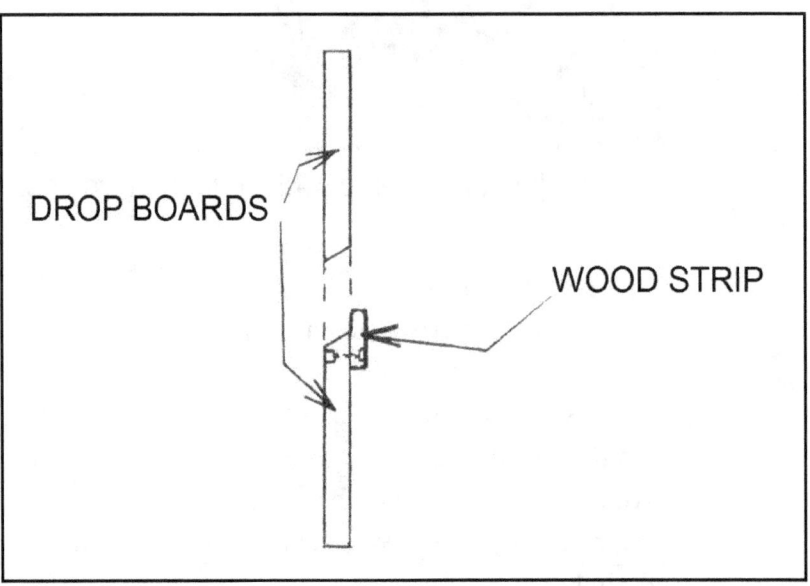

Adding wood strip to reinforce drop board.

DROP BOARDS

WOOD STRIP

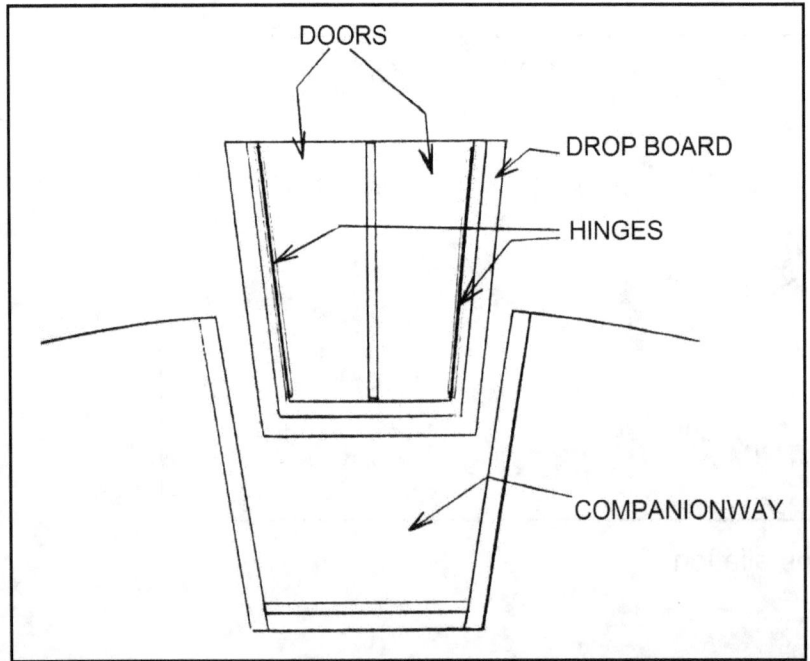

DOORS

DROP BOARD

HINGES

Doors installed in one piece drop board.

COMPANIONWAY

Other possible modifications include changing from drop boards to doors, or from doors to drop boards. Doors can also be installed in a one-piece drop board, as shown. This arrangement allows the more convenient doors to be used at dockside and the more secure drop boards to be used when the boat is under sail or power in unprotected waters.

Ventilators

Many types and sizes of ventilators are now on the market. The basic idea is to let air in or draw it out while preventing water from getting below. Dorade-type and baffled-cowl ventilators are both popular and effective. Manufactured models come with mounting instructions, which usually makes installation fairly easy.

Dorade ventilator.

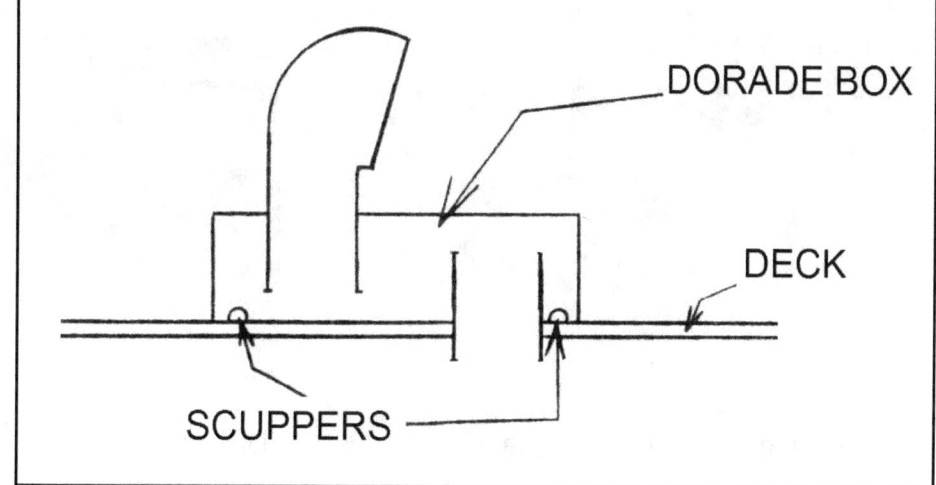

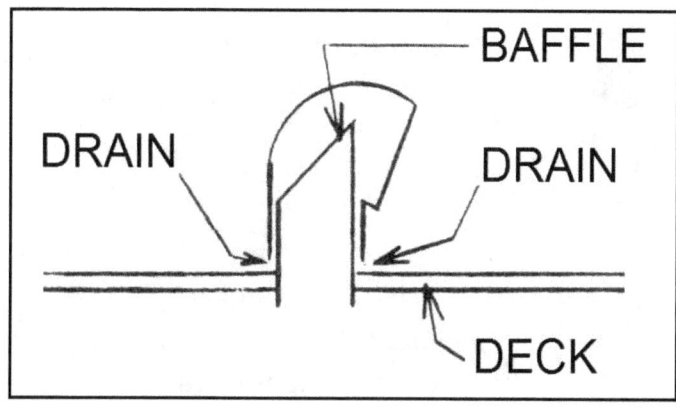

Baffled-cowl ventilator.

Another popular ventilator is the Tannoy, which is made of stainless steel and plastic. I installed one of these on the forward hatch of a Westerly sailboat.

Tannoy ventilator.

Sawing hole with saber saw.

Drilling hole.

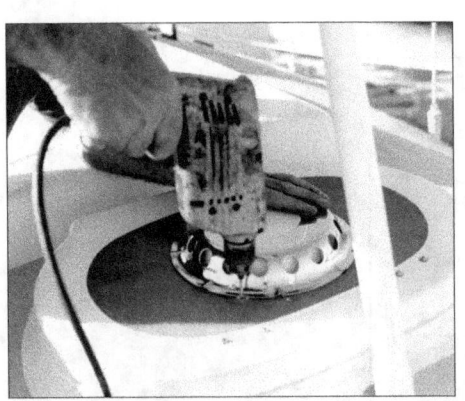

Drilling hole.

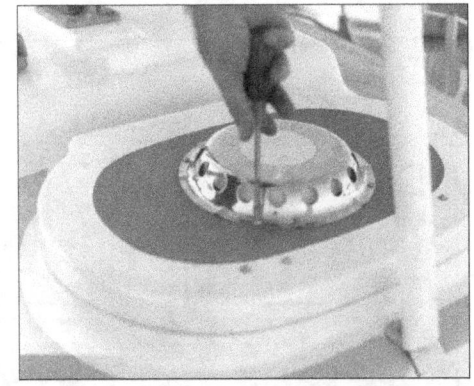

Installing fasteners.

11

INTERIOR ACCOMMODATIONS

The interior of a boat is almost certain to be a series of compromises. With limited space, a larger something means a smaller something else. In planning interior modifications, keep this in mind. The interior of stock boats are designed for an average buyer. In actuality, this person does not seem to exist.

Typical complaints are these: the boat has too many berths, it doesn't have enough storage space, it's uncomfortable, the inside looks more like an icebox than a boat, and it just doesn't work at sea. In many production boats, for example, cushions are thrown all over the cabin in a moderate sea, doors fly open, sharp corners are everywhere, and there's nothing to hold onto.

There are many possible modifications that can be made to improve the interior of a boat, and many of the jobs are fairly easy. In this chapter, a number of typical modifications are detailed.

Dinettes
In some boats, a berth can be converted to a combination dinette/berth. A typical conversion is shown. The sections of plywood can be bonded with fiberglass strips to a fiberglass molding. A soft-core material should be used between the wood and all connections to the hull shell. Bonding methods are similar to those used in installing a plywood bulkhead, except that the fiberglass bonds generally need not be as thick. If the berth was originally constructed of plywood, the added plywood sections can be joined by glue and screws to a common frame piece.

The table top can be constructed in any of several ways. One way is to use plywood covered with a plastic laminate as a table top.

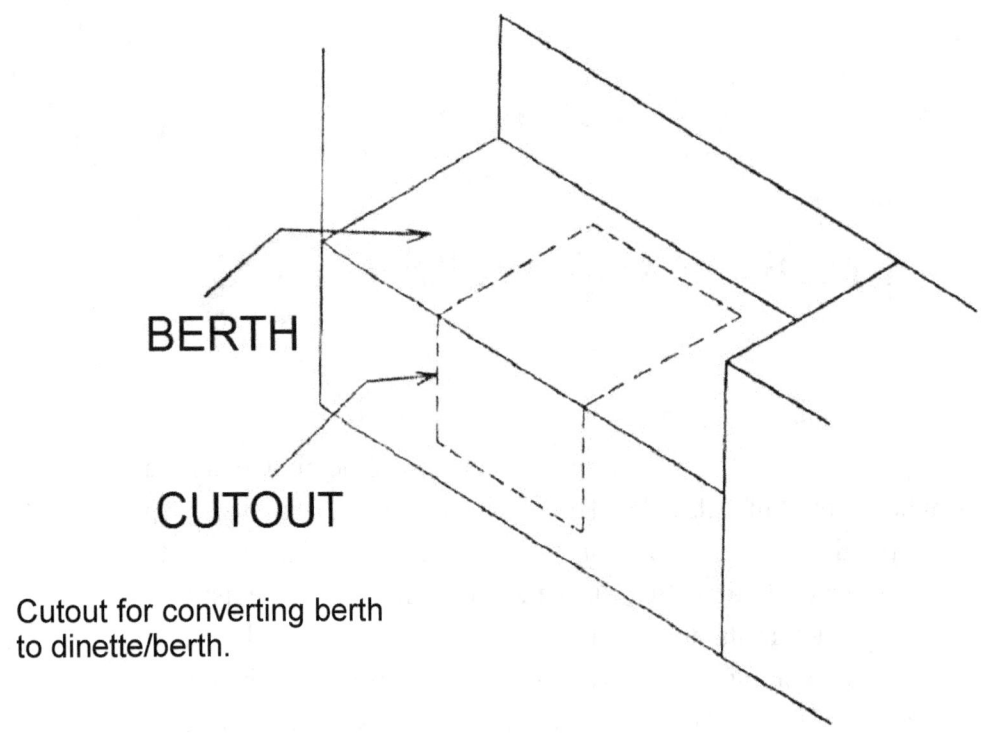

BERTH

CUTOUT

Cutout for converting berth
to dinette/berth.

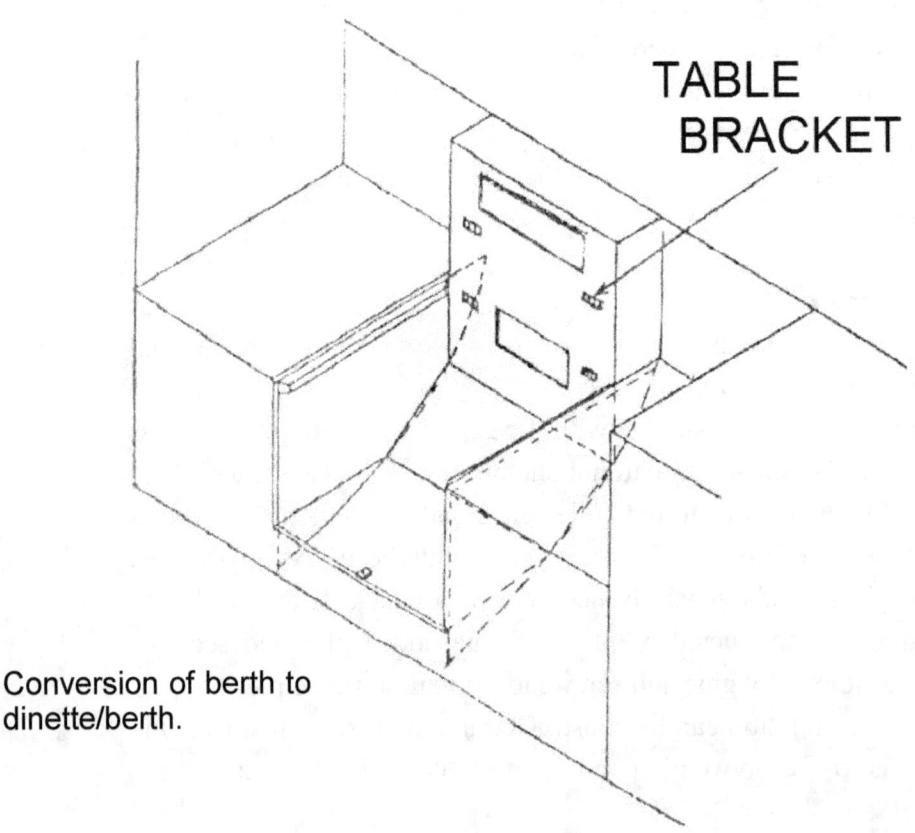

**TABLE
BRACKET**

Conversion of berth to
dinette/berth.

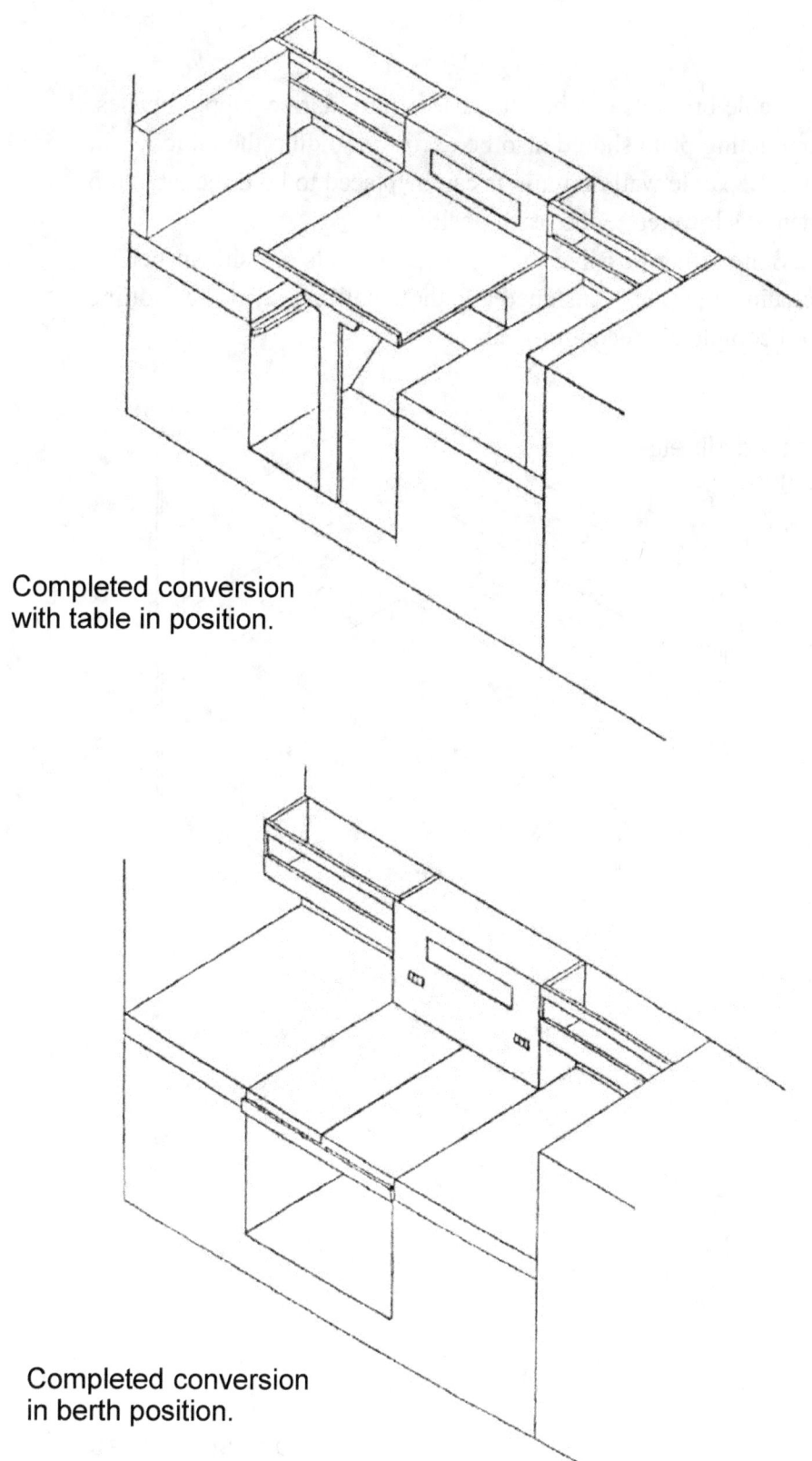

Completed conversion
with table in position.

Completed conversion
in berth position.

Table brackets can be purchased from marine supply houses. A mounting plate should also be used for holding the table leg in place. Separate wall mountings can be placed to hold the table top when it is lowered to the berth position.

Some other possible dinette arrangements are shown below, including the one I designed for the small cabin in the Dotline that I completely reconditioned.

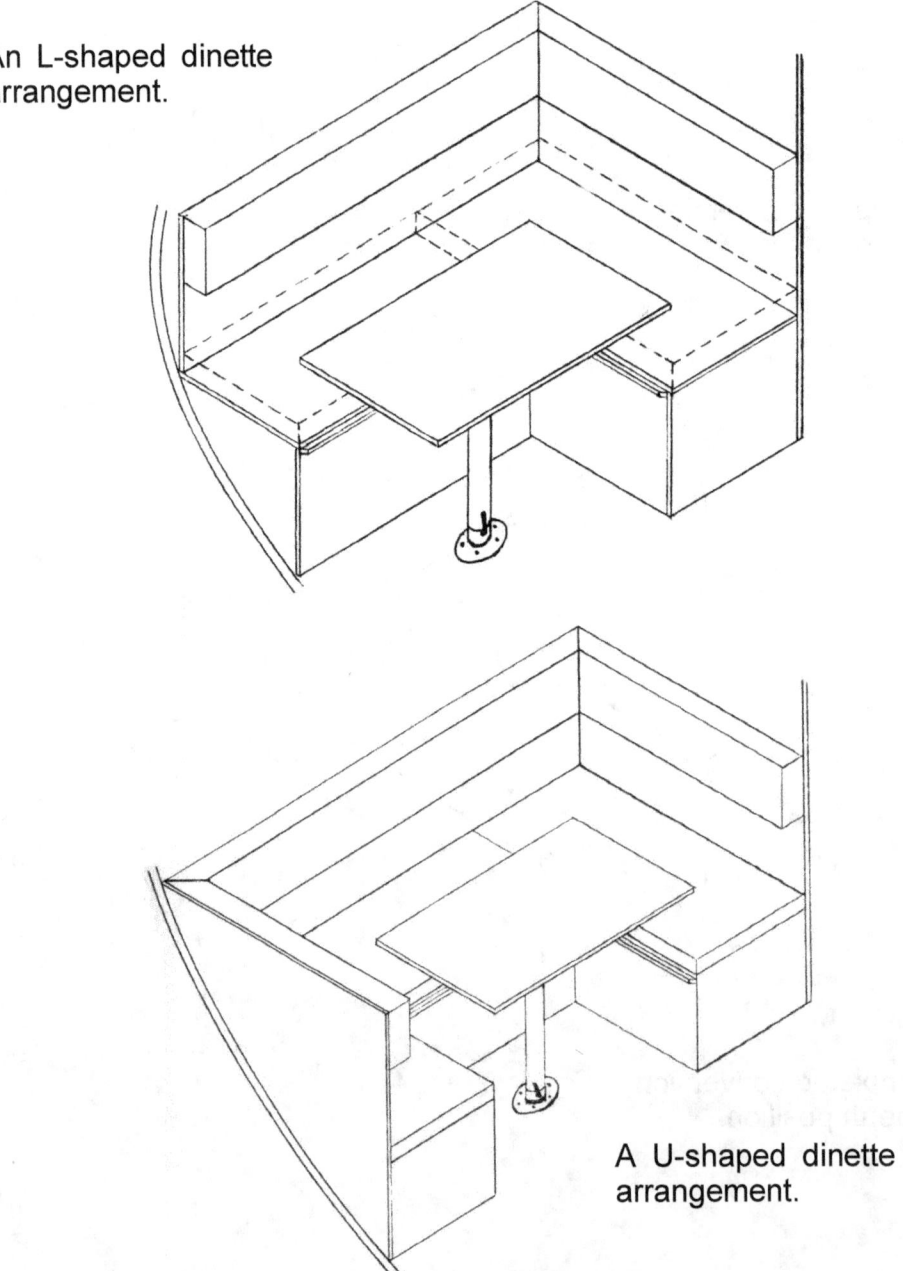

An L-shaped dinette arrangement.

A U-shaped dinette arrangement.

Tabletop lowered to berth position during construction on Dotline.

Seat for table during construction on Dotline.

Table in use on completed Dotline.

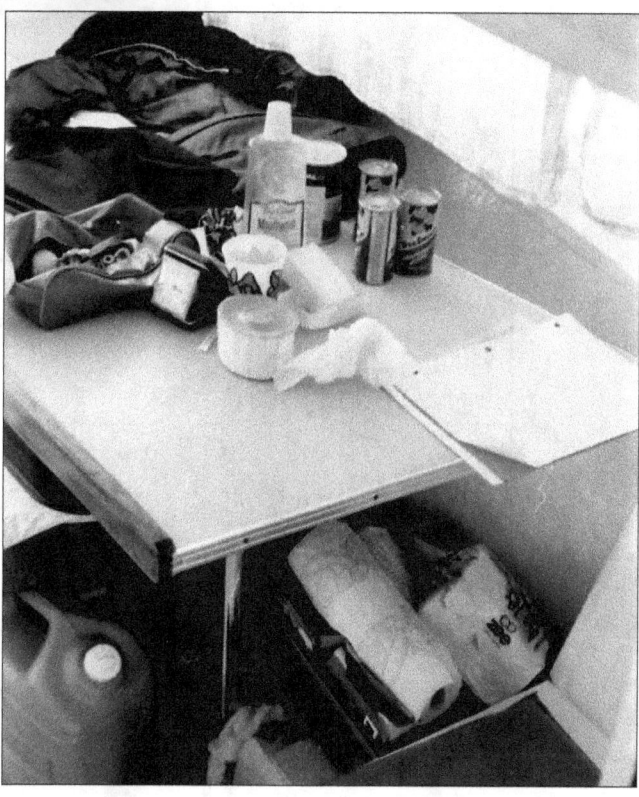

Folding Tables

A folding table between two berths is a popular arrangement for offshore cruising. A typical construction is shown. The table top should be approximately 12 inches above the top of the settee berth, and, thus, about 27 inches above the floor.

Standard marine hardware can be used for brackets; or the brackets can be constructed from wood and hinges. The table should be mounted solidly to the cabin sole, as considerable stresses and strains will be applied to the table. The shelf provides effective bracing.

The types of wood to use and the size of the table top are pretty much up to the builder, except that the sides of the table should be approximately even with the edges of the settees. Sea rails can be added to the table top.

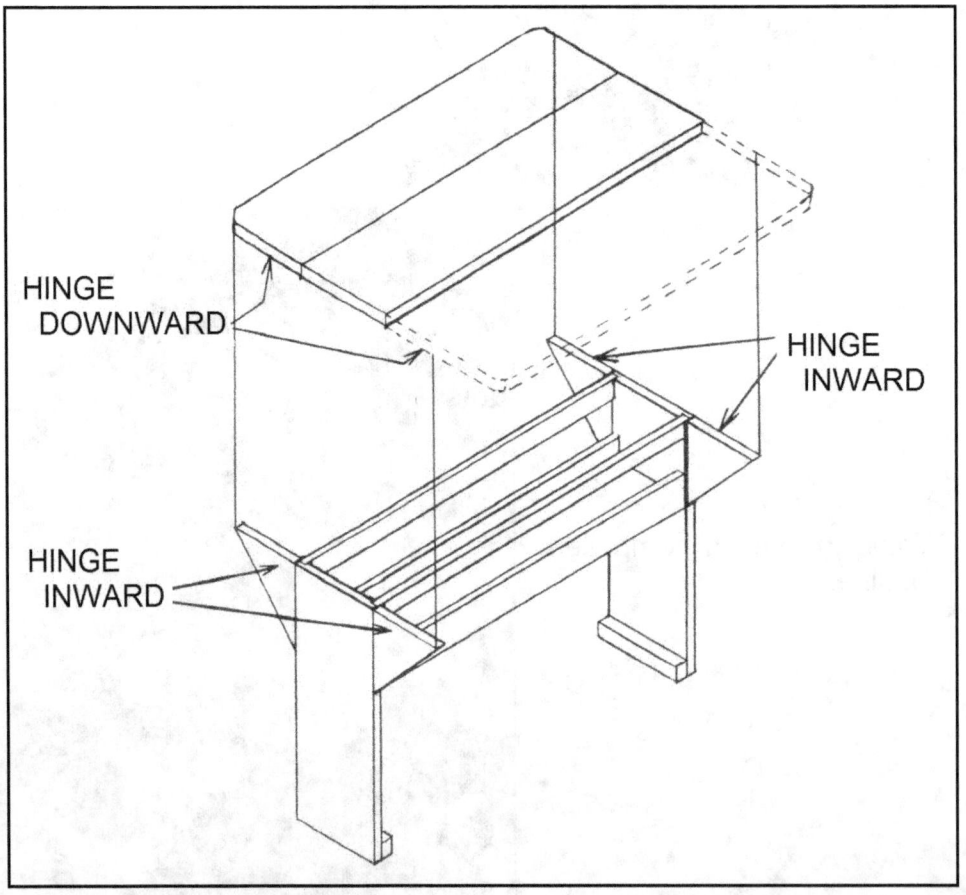

Construction of a folding table.

Openings to Lockers and Cabinets

A simple cutout in the fiberglass panel can be used for access to a storage compartment. After the opening is cut, the exposed fiberglass laminate should be coated with resin. The opening can also be lined with a wooden frame, as shown.

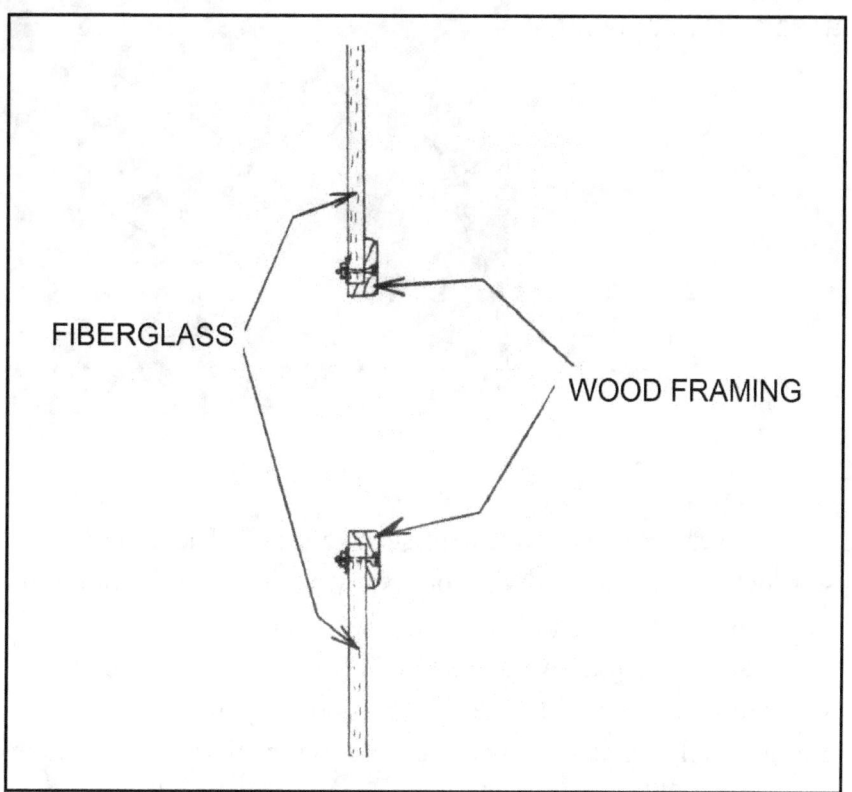

Wood framing around locker opening in fiberglass panel.

Sliding doors can be added by fastening double channels to the wood framing. Doors can be of wood, acrylic plastic, or other suitable material.

Doors with hinges can also be used. The hinges should be through-bolted to thin plywood and fiberglass laminates. Several types of latches are now on the market.

Access openings can be made under berth cushions. A cover can be made from plywood.

Construction of hatches on Dotline for access to spaces below berths.

Drawers

Drawers provide convenient storage space. These should be self-locking so that they will not open when the boat is rolling in a seaway.

Manufactured drawers are available. These can be bought in standard sizes, or they can be custom-made to your patterns. They are generally expensive, but are an alternative for someone without the skill and/or equipment for constructing them. Some drawers come with a sliding track for easy mounting.

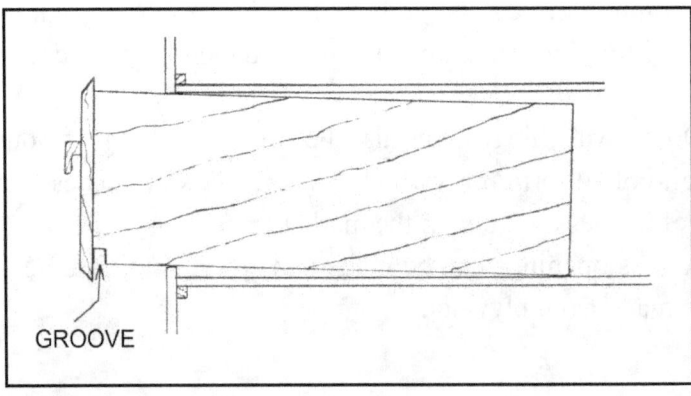

Self-locking drawer.

Berths

Many stock fiberglass boats are crammed with berths, but often none of them are comfortable or safe. Many owners desire to convert from a number of inadequate berths to a few good ones.

Size is an important consideration. It has been my experience that a berth must be at least 8 or 10 inches longer than the height of the user to be comfortable. This means that most berths should be at least 6½ feet long. Seven feet is even better.

Opinions vary on the width of berths. For an offshore boat, some experienced sailors recommend a narrow berth, in some cases, less than 2 feet wide. I feel that 28 to 30 inches is better, provided that adequate lee canvases or bunk boards are used. If necessary, pillows or life jackets can be used as wedges in rough weather. This way, a wider berth is available for use in hot weather, under calm conditions, and in port.

Double berths are nice in port, but are generally unsuitable at sea except in calm conditions. One acceptable arrangement is to have the mattress in two sections, with a wooden bunk board or lee canvas that can be attached between the mattress sections for use in rough weather.

Being able to sit up in a berth is generally considered desirable. Having slept – or attempted to anyway – in berths in trimarans and in pilot berths in monohulls with only a couple of feet between the top of the cushion and the deck or cabin overhead, I agree with this. It's also nice to have something at the head end of the berth to lean back on when sitting.

Adding a removable piece of plywood between V-berths is a simple way to form a double berth. A separate V-cushion can be fitted.

Bunk cushions on many stock boats are only 2 or 3 inches thick. Many owners add an inch or two of padding. You can have this done at an upholstery shop or, if you have the necessary skills, do it yourself. The padding should be polyfoam or some other material that will not absorb water. Vivatex canvas covers can be used in place of Naugahyde. Plastic covers are often too hot for comfortable sleeping, although they are easy to keep clean. Cloth covers should have a zipper on one side so that they can be removed for cleaning.

Berths on Dotline under construction.

Shaping foam for cushions.

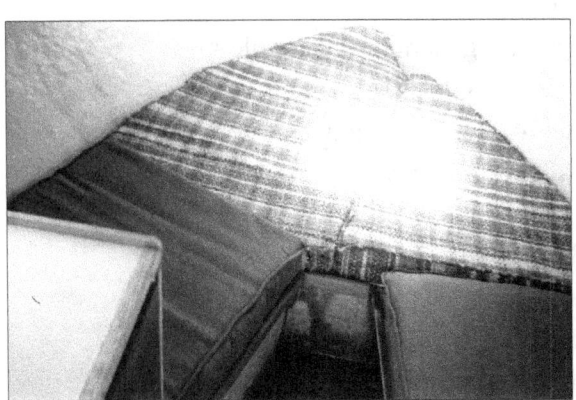

Completed cushions.

Berth in use on completed Dotline.

Many berths are also used as settees. These should have a padded backrest for comfort. Removable backrests in the form of cushions or a padded board can be fitted.

If the berths do not have rails to keep the cushions from sliding off, these can be added. These are generally about 3/4 inch above the bottom board of the berth.

Galleys

Depending on the tastes of the owners, galleys occupy too little or too much of the interior space. One owner may wish to subtract from the galley area to add a navigation area or a wet locker, but another owner will gladly sacrifice a bunk to have space for a refrigerator or a larger galley counter.

A fairly frequent modification is to change the location of the galley. This job can range from a minor to a major modification, depending on the boat and how the galley is to be constructed. If plumbing has to be relocated, especially through-hull fittings, the difficulties increase. The cost of this modification depends to a large extent on how much of the old galley materials and equipment can be used for the new one.

A common modification is to add gimbals to the present galley stove or to change to another stove. Each person seems to have his own preference for the fuel to be used. Coal, wood, alcohol, kerosene, diesel, and bottled gas (propane) all have their advocates. Even electric stoves are popular: power is available both at the marina and, if appropriate, generators are used away from the dock.

Stoves should be fastened securely in place or to gimbals. Small stoves without ovens are frequently through-bolted directly to the galley counter or connected to gimbals attached to the counter. Small gimbaled stoves can also be set in a shallow recess. Some small gimbaled stoves come with a mounting frame, which can be mounted flush in a cutout in the galley counter top. Make sure there is ample clearance for the stove to pivot without hitting anything at the sides.

Stoves with ovens are frequently mounted on gimbals to partial bulkheads in a deep recess. Stove makers will, on request,

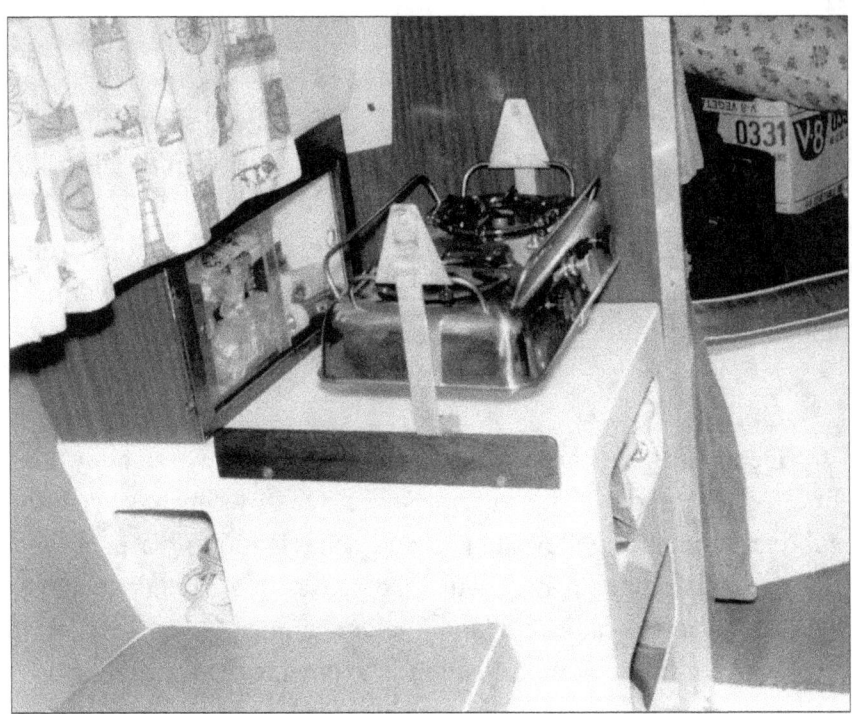

Gimbals mounted on counter top.

send diagrams showing the space requirements for their stoves. Also, these measurements are frequently shown in marine catalogues.

If bottled gas is used, the container should be mounted on deck or in a vapor proof locker that is vented overboard. Opinions vary about the use of bottled gas on boats. I feel that the risk is too great for the added convenience. However, if bottled gas is used, there should be a shutoff valve right next to the bottle. Whenever the stove is not being used, this valve should be closed.

Galley counters can be constructed as desired. One method is to cover plywood with a plastic laminate, such as Formica. The installation should be such that water cannot leak through joints at the edges or around a sink mounting. If fiddles are mounted, they should be sealed to the counter top.

Sinks are another important consideration. You may want to change from a fiberglass-molded sink to one of stainless steel or to change to a larger and/or deeper sink. Stainless steel sinks can be purchased in many sizes and shapes. These generally have fastenings by which they are mounted in a cutout in a counter. To

Starting Dotline galley modifications.

Icebox below counter.

Sink installed in countertop.

Water pump installed.

Galley construction.

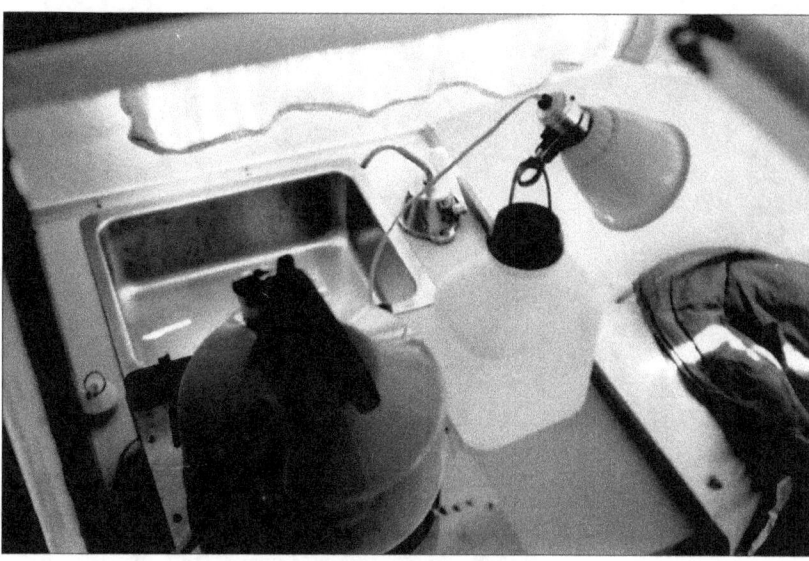

Completed galley in use with Dotline afloat.

install a sink, first make a template of the cutout if one is not supplied with the sink. The tolerance should be as close to a perfect fit as possible. Mark the cutout and saw it out. After a dry fit has been achieved, apply bedding compound and fasten the sink in place.

A place for everything applies in particular to the galley. Storage compartments should be placed so that needed items are close at hand.

Iceboxes and refrigeration units are frequently installed in boats. Manufactured units complete with mounting instructions and hardware are available.

Toilet Compartments

A typical modification is to add a holding or recirculating tank. In many areas, there is still considerable confusion as to what constitutes a legal system. Marine stores are generally an unreliable guide to what is needed, as they are generally more interested in selling the higher priced units. It is often helpful to find out what systems other boaters in your area are using.

If space permits on your boat, a counter, sink, and storage compartments can be added to the toilet compartment. A shower is another possibility.

12

PLUMBING

There are many possible plumbing installations and modifications that can be made on fiberglass boats. This is a specialized subject, and space permits only a sampling here.

Through-Hull fittings and Sea Cocks

The standard mounting of a through-hull fitting is shown. If you have never installed a through-hull fitting, practice installing one with scrap material.

The basic steps are as follows:

1. If required, reinforce the hull with additional layers of fiberglass.

2. Shape a wooden backing block and epoxy-glue it in place to the inside of the hull.

3. Drill a close-tolerance hole.

4. After a dry fit has been achieved, apply a bedding compound, install the fitting, and tighten the nut.

5. Remove excess bedding compound.

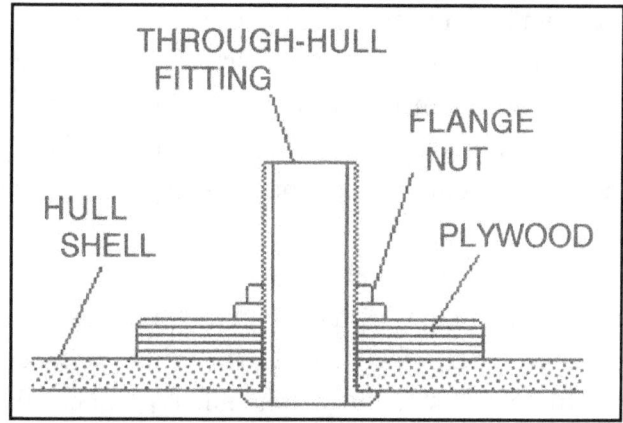

Standard mounting of through-hull fitting.

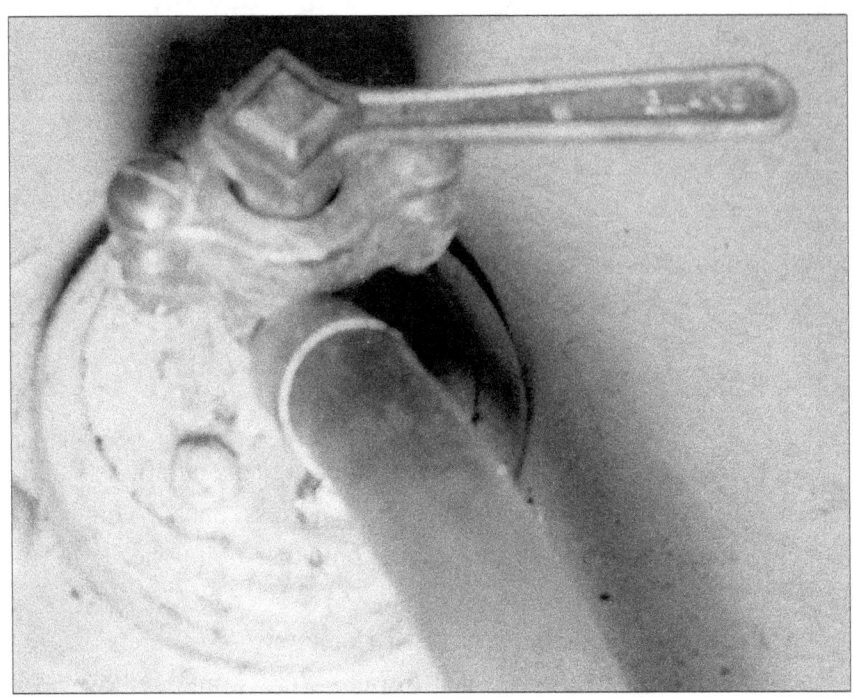

A sea cock.

As a general rule, through-hull fittings below or near the waterline should have sea cocks.

Bilge Pumps

On many stock fiberglass boats, bilge pumps are installed only as an optional extra. It's surprising how many beginning boaters do without them, yet they are among the most important safety equipment on a boat. Further, every boat that doesn't have positive buoyancy or that is used offshore should have at least one, and preferably two, *mechanical* bilge pumps. These should be operable from both the cockpit and inside the cabin.

Electric bilge pumps are a great convenience, but they should always supplement the mechanical pumps, not replace them. The electrical system is most likely to fail when the bilge pump is most needed – in an emergency. Some bilge pumps operate from the engine.

In addition to an electrical pump, there should be at least one hand pump. The two types of hand pumps used most often today are the plunger type, which should be of large capacity and constructed of metal, and the diaphragm type. The diaphragm type has the advantage of double action – that is, water can be pumped when the handle is moved in either direction. Some models can be mounted below deck so that the pump can be operated both there and above deck. Dual handle attachments and a watertight deck socket make this possible.

Regardless of the type of hand pump used, it should be mounted so that it can be operated from the cockpit and/or inside the cabin. Though the pump itself can be located below deck or inside a locker, the pump handle should be operable from the cockpit without having to open a locker, especially one that drains to the bilge.

A typical bilge-pump mounting is shown. This below-deck method eliminates having hoses pass through the deck or the sides of the cockpit. The pump should be fastened in place with through bolts.

The hoses should be of large diameter, generally large enough to fit over the inlet and outlet of the pump, and should be non-collapsible. Generally, the inlet is placed in the bilge sump,

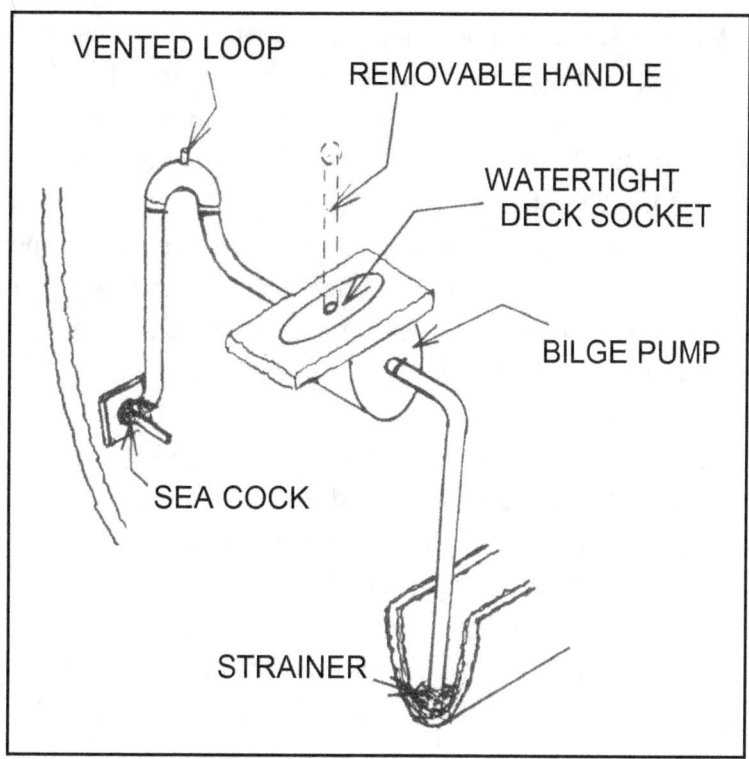

Typical bilge pump mounting.

and has a long enough hose that it can also be used in other locations. A strainer should be used on the inlet to the hose.

The outlet should pass through the hull well above the waterline, if possible, so that the outlet will not be submerged when the boat heels. If this is not possible, the hose should be looped and vented.

Stainless steel hose clamps should be used throughout the system.

Other possible modifications to bilge pumps include installing dual-pump suctions fitted to a Y-valve so that water can be pumped from either of two locations, and making it possible to pump sea water through a hose for washing down the deck. However, remember that the latter modification is limited to the area where the pump handle can be operated if the hose is to be held by the same person who does the pumping.

A number of electric bilge pumps are on the market. These are installed like hand pumps. Follow the manufacturer's mounting instructions, especially in regard to wiring.

Water Systems

Water systems in stock fiberglass boats range from simple hand pumps connected to plastic water jugs by a length of hose to large, built-in tanks with pressure systems. Another possibility is a gravity system. However, this system is not popular today, nor is it found in many stock fiberglass boats.

In most systems, the water is drawn from the tank by means of a hand pump. A frequent modification is to add foot pumps, which make such tasks as washing hands much easier. Some owners want to change from hand or foot pumping to electric pressure systems.

Hand and foot pumps are readily available and easy to install. Among them there is considerable difference in quality and ease of operation, however. Though good plastic pumps can be made, many of the cheaper ones do not last very long. Many of the better ones are made of both stainless steel and plastic.

Hand pumps are installed in the line between the tank and the waterspout. Some are attached only between two flexible hoses;

others are also fastened to the floor. The waterspout is installed in a manner similar to that used for hand pumps with the waterspout attached.

A number of units are available for converting to pressure systems. A faucet is used, and the pressure unit (water pump) is installed in the water line between the faucet and the tank and is wired to the battery. The better units are completely automatic and draw current only when the faucet is opened. There are two disadvantages to this system: water consumption is increased; and if the battery goes dead, so does the water system. To avoid this, many owners also install a hand-pump system, which conserves water on a long cruise and can be used if the pressure system fails.

Many owners also install a sea-water system to the galley sink. A typical installation is shown. Since the inlet is well below the water line, a sea cock should be fitted. In some cases, an existing sea-water inlet can also be used for the galley. This reduces the number of through-hull fittings below the water line.

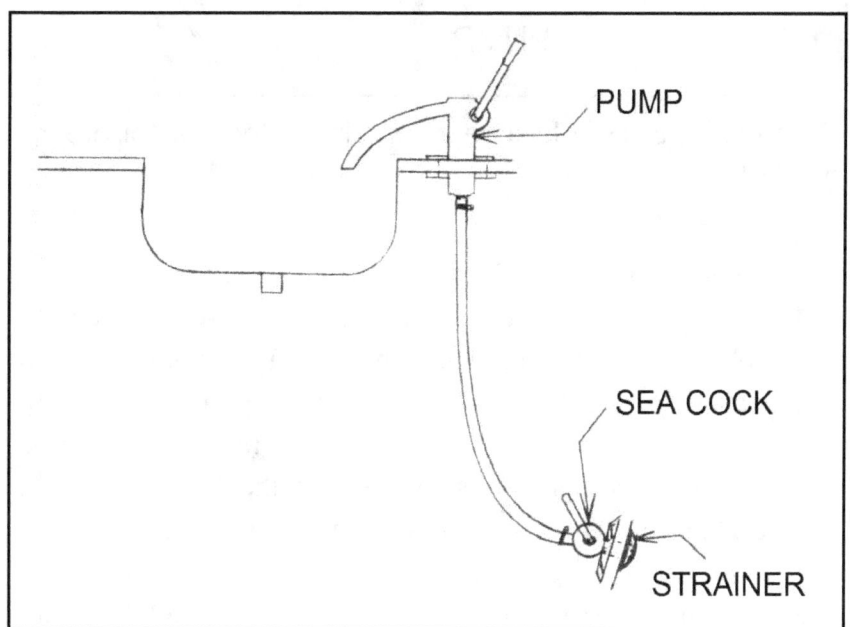

Sea-water system to galley sink.

If the sink will always be above the water line, even when the boat is at its maximum angle of heel, the drain can be strictly a gravity system. The through hull for this is generally near the water line. A sea cock should be used.

If the sink is too low in the boat for a gravity drain system, the pump system shown below can be used. The through-hull outlet should be well above the water line. If this is not possible, a vented loop should be used in the line to prevent back-siphoning.

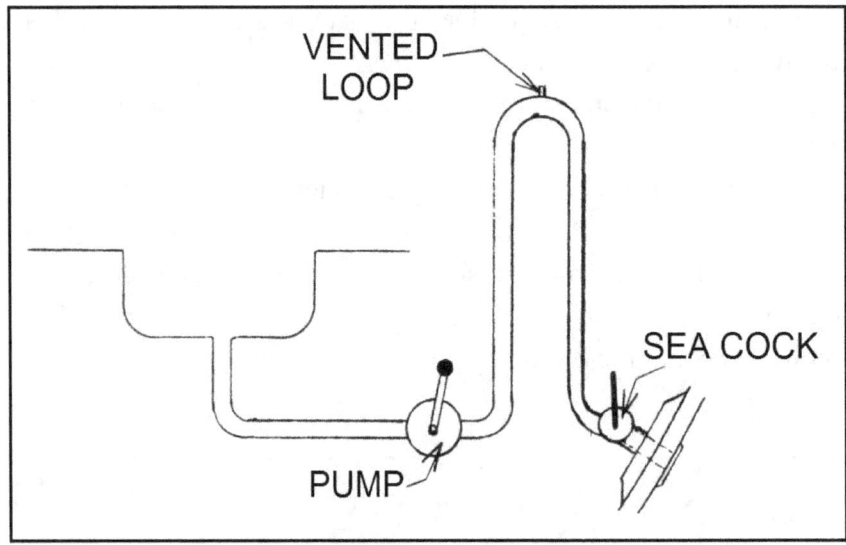

Pump drain system for use when sink is too low for gravity system.

Marine Toilets

A common modification is to convert a direct-discharge head to a holding-tank system, which may or may not be of the recirculating type. Before purchasing a holding tank, make sure that it meets the regulations in the areas where the boat will be operated. In most cases, this means that the holding tank is connected to a deck plate for hookup to a disposal unit.

Some marine toilets have the holding tank built in, which makes installation easy. These are best purchased as a kit with all necessary hardware for installing the unit on a boat.

13

ELECTRICAL SYSTEMS

The design of electrical systems is a complex subject that is beyond the scope of this book. Most modifications involve adding to the present electrical system, adding a basic electrical system where none was installed previously, and adding connectors for shore power. It is these jobs that are detailed here.

A Basic Battery System

In installing a basic battery system, I recommend that a wiring harness that meets marine safety standards be used. Several manufacturers make these. The terminals are marked for easy installation. A typical unit for a runabout includes a fused or circuit-breaker panel with three switches, which can be used for running lights, bilge pump, bilge blower, spotlight, or other electrical accessories; the wires are made up in a harness and color-coded. A wiring kit for a cruising boat might include five switches on the panel with fuses or circuit breakers for running lights, anchor light, cabin interior lights, bilge pump, blower, and other electrical accessories, as well as a prewired harness that is color-coded. Make sure the wiring harness and panel is compatible with the voltage you intend to use. Most lights and accessories operate on 12 volts.

Three principles should be followed in wiring a boat:

1. The wire size should be large enough to carry the required current with a minimum voltage drop.

2. Insulation must be adequate for the line voltage under marine environmental conditions.

3. Dangerous areas must be avoided, and the wires must be fastened securely in place and protected from vibration and other forces.

By using an approved wiring kit, the first two principles are taken care of. Locating and fastening wires needs further elaboration.

Wires are generally installed in out-of-the-way areas, such as under berths and inside lockers. However, they should be above the bilge level. The wires should have mechanical fasteners spaced not more than 14 inches apart. In addition, the wires should be protected at points where mechanical damage is likely. Drain holes should be used whenever the wires run inside a tube, so that trapped moisture will have a means of escape. When wires must pass through the hull shell or a watertight bulkhead, special through fittings, available from marine stores, should be used. These usually consist of a socket and plug; the socket is generally mounted. Holes are drilled for the wire and mounting bolts. Bedding compound is applied, and the socket is bolted in place.

Approved wire clamps, available from marine stores, should be used. These are generally held in place with one or two screws. Double-pointed staples should not be used, as these are likely to damage the wire insulation and cause a short circuit.

Opinions vary on whether crimp-on connectors or soldered wire joins should be used. If crimp-on fittings are used, these should be Underwriters Laboratories-approved pressure connector lugs. If soldering is used, the solder should be a high grade of resin-core electronic solder.

Assuming that the wires in the harness are of the correct length, all connections will already have been made. If the lengths of any wires are extended, either a soldered or a crimp-on connection can be made. A special crimping tool is needed for installing the lug connector. Electrical tape can be added. Exposed wires are subject to rapid corrosion.

Since wires carrying current are surrounded by a magnetic field, they should be kept away from the compass. Where this is not possible, as with the wires to the compass light, the wires should be twisted together to reduce the magnetic effects.

If any modifications are made to the wiring harness, it's suggested that the same wire sizes and types be used, and that the same color-code system be continued. This makes trouble-shooting much easier.

If any additional circuits are to be added to the present system, try to obtain the wiring diagram from the manufacturer. Also, follow carefully the wiring instructions for the electrical accessory to be connected, especially in regard to wire size and amperes for fuses or circuit breakers. Hopefully, the panel will have space for additional circuits.

As a general rule, the negative lead (assuming a negative ground system) should run continuously from the electrical accessory to the negative terminal of the battery with no fuses, switches, or circuit breakers in between. This will make sure that there is no potential difference to the ground when the accessory is turned off. The positive lead generally runs from the electrical accessory to one terminal of the fuse or circuit breaker, and then continues from the opposite terminal to the positive terminal of the battery. Switches are generally placed in the positive lead at the electrical accessory and between the battery and the fuse or circuit breaker.

Battery Location and Installation

If the boat has a battery for the engine, this is generally also used for electrical accessories. Sometimes a dual-battery system is used. In most systems, both batteries are charged by the alternator or generator of the engine. However, one battery is generally used primarily for starting the engine, and the other for electrical accessories. A switching arrangement is usually included so that either battery can be used for starting the engine in an emergency.

If the battery is already installed, connecting a wiring harness kit is easy. Without an engine with an alternator or generator, some other means is needed for charging the battery. Shore power and an automatic battery charger or an auxiliary generator is commonly used. With the shore arrangement, the time away from the dock before the batteries will run down is limited. The time varies, depending on the battery, electrical accessories used, and so on, but in most cases, the battery will last 2 or 3 days before running down.

The battery should be located in a well-ventilated area, as the lead acid battery (the kind commonly used in boats) gives off hydrogen when it is being charged. If ventilation is poor, there is the danger that hydrogen may collect and ignite. The battery should also be located above bilge level, and should be secured in a battery box. Opinions vary as to what type of box should be used. Plastic boxes with lids are found on many boats, but some experts feel that these do not allow adequate ventilation to the battery. A frame case with a pan below the battery allows ventilation to reach the battery, yet holds the battery securely in place and catches any dripping battery acid.

Regardless of the type of battery box used, it should be securely mounted. A simple way of doing this is to bond the box to a bulkhead or, with a pad in between, to the hull shell.

Ground Systems

Marine electrical engineers do not agree on what ground system should be used on boats. Perhaps the most commonly used method is bonding. A heavy copper strip or wire is run the length of the boat and connected to the negative terminal of the battery. Everything that requires grounding – such as metal tanks, radio ground plates, the engine block, and sometimes chain plates (for lightning protection) and through-hull fittings if they are metal – is then connected to this. The copper strip or wire is connected to a ground plate, which is located outside the hull in the water.

If this method is used on the boat, then all additional metal components added to the boat should be similarly connected to the grounding system. In any case, the wiring diagram of the system presently on the boat should show the ground system used.

Lights

The most common electrical accessories on boats are lights. A wide range of light fixtures is manufactured. For interior lights, standard, fluorescent, and LED lights can be used. The latter have the advantage of drawing less amperage and giving off less heat.

Interior lights are mounted in various ways, such as by fastening a plate to a bulkhead or table top. In most cases, each light has an individual switch, besides having a circuit switch on the main panel.

Typical modifications include adding lights where none were previously installed, relocating lights, and changing to different fixtures.

Many types and designs of exterior lights, such as running, anchor, and deck lights, are on the market. These vary greatly in quality and suitability. The kind with a cap held in place with screws tapped into the fiberglass shell has proved to be totally inadequate, yet many stock fiberglass boats come equipped with these.

In some cases, a through-hull watertight socket is part of the light fixture. This arrangement allows changing the lens and bulb without breaking the through-hull seal, as separate hull and lens fasteners are used. The light fixture should, in most cases, be through-bolted to the hull. In most cases, large washers are adequate for backing plates.

Another type of exterior light is mounted outside the boat, such as on a pulpit, and a separate watertight socket is used for passing the wire through the hull. Passing a wire through a small hole in the hull without the special socket is not recommended. As leaking is likely even if bedding compound is used. The wiring on deck should be placed so that no one will trip over it.

Installing mast lights involves additional problems. Though many lights have been installed on masts while the masts are standing, it is generally easier to lower the mast. Typical locations are on the masthead, forward of the mast just above the spreaders, and on the spreaders. The lighting requirements vary with the size of the vessel and where it is to be used (inland or international). If you are in doubt as to where the boat will be used, you should install international lights, as they are legal for both inland and international waters.

Light fittings typically come with the fasteners by which they are attached to the mast. Running wires through the inside of a hollow aluminum or wood mast often presents difficulties. One way to do this is to first run a piece of stiff baling wire from the mounting hole for the light fixture to the bottom of the mast. The electrical wire is then attached to this and pulled through.

One arrangement for the wire at the bottom of the mast is to run it through a watertight socket to the outside of the mast just above deck level, and then through the deck with a watertight socket and plug. This method is especially suitable for hinged masts.

If a non-hinged mast heel is used, the wires are sometimes passed through the deck inside the mast. If the mast is stepped on the keel, the typical arrangement is to run the wires out of the mast below deck level.

Other Electrical Accessories

In addition to the lights, many other electrical accessories can be run off the battery, such as horns, cigarette lighters, windshield wipers, bilge and water-pressure pumps, blowers, alarm systems, refrigerators (these are often run on AC from the engine alternator, shore power, or by stepping up the battery voltage by means of a solid-state inverter), electric toilets, windlasses, capstans, power winches, and an endless array of electronic equipment. The limit seems only to be the capacity of the batteries to be recharged.

The wiring instructions should be carefully adhered to when installing electrical equipment, especially in regard to grounding and the amperage of fuses and circuit breakers.

Shore Power

Wiring methods for using shore power aboard a boat vary. Perhaps the most common system is to use a bonding system to the boat's grounding plate – which should be a copper strip or wire separate from that used for the battery system – in addition

Shore power electrical hookup and battery charger.

to the shore grounding wire. The wiring for shore power should be a separate system, isolated from the battery wiring.

Wiring for shore poser should never be a haphazard affair. There is danger not only of electrolysis but also of electrocution.

Through-hull inlet fixtures with watertight covers are commonly used for connecting shore power to the boat wiring system.

Electrolysis

One method of keeping electrolysis to a minimum is to use only one kind of metal in the boat. However, this is generally impractical; even if it is done, it does not prevent electrolysis that results from metal on other boats moored nearby.

A number of induced-current systems for electrolysis prevention are now on the market. These systems are somewhat complicated and require careful adjustment, but some users have reported that they work well.

There are varying opinions as to using a common bonding system for preventing electrolysis, as the results are not always predictable.

The most commonly used preventive of electrolysis is probably zinc plates, which serve as sacrificial anodes. These plates are attached to the hull near or on the metal fittings. If not attached directly to the fitting, the zinc should be connected to it by a copper strap. Zinc plates come in a variety of sizes and shapes. One type fits propeller shafts.

Since zinc is eaten away, it must be replaced from time to time, generally after 10 percent of it has corroded.

14

SPARS AND RIGGING

Most stock fiberglass sailboats now come equipped with aluminum masts and booms and stainless-steel standing rigging. A frequently desired modification is to beef up the rigging. Other common modifications include adding reefing and furling systems, blocks, downhauls, chafing gear, and gallows frames.

Chain Plates

Chain plates are used to connect the standing rigging to the hull. Modifications include replacing present chain plates with stronger ones and adding additional ones. The ease with which chain plates can be removed and replaced with stronger ones varies greatly from boat to boat. It's almost always a major job, so the need for this modification should be carefully considered.

The chain plates on most stock fiberglass boats being manufactured today are mounted inside the hull. A present trend is to mount the chain plates for the shrouds to bulkheads. This type of chain plate is generally one of the easiest to replace with a stronger one. Bolting chain plates through the hull shell is being done on fewer and fewer manufactured boats. A number of manufacturers are now bonding chain plates to the hull. Strands of glass fibers pass through holes or around pins in the chain plates, which are often upside-down V- or fan-shaped to spread the load over a wider area of the hull. These glass fibers are wetted out with resin and then bonded to the shell. Several reinforcing fiberglass layers are generally added over this laminate.

Some sizes and shapes of chain plates can be purchased at marine stores. Others must be made at machine ships.

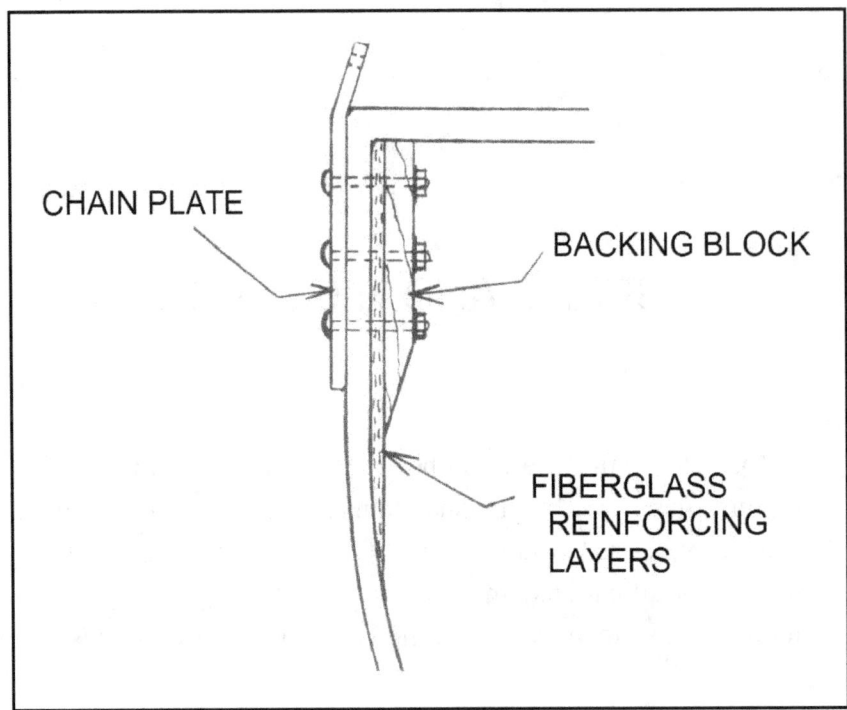

CHAIN PLATE

BACKING BLOCK

FIBERGLASS
REINFORCING
LAYERS

Method for mounting chain plate outside hull.

Probably the easiest method for the amateur builder is to mount chain plates outside the hull. This method is used quite often in beefing up a stock fiberglass boat for an ocean passage.

The inside of the hull in the areas where chain plates are to be mounted should be heavily reinforced with additional layers of fiberglass and a wood or metal backing.

The chain plates can be square-punched for carriage bolts, or round holes for round bolts can be drilled in the chain plates. Mark the location of the chain plates and the positions of the holes for the fasteners. The holes through the fiberglass shell should be a tight fit for the bolts. Stainless-steel bolts should be used for stainless-steel chain plates.

After a dry fit has been achieved, the contact area of the chain plate on the outside of the hull should be sanded. An epoxy adhesive can be used between the chain plate and the hull before the final fastening. Bedding compound is used on the fastenings.

The inside of the hull is reinforced in a similar manner for chain plates that are mounted internally. The bolts that are used

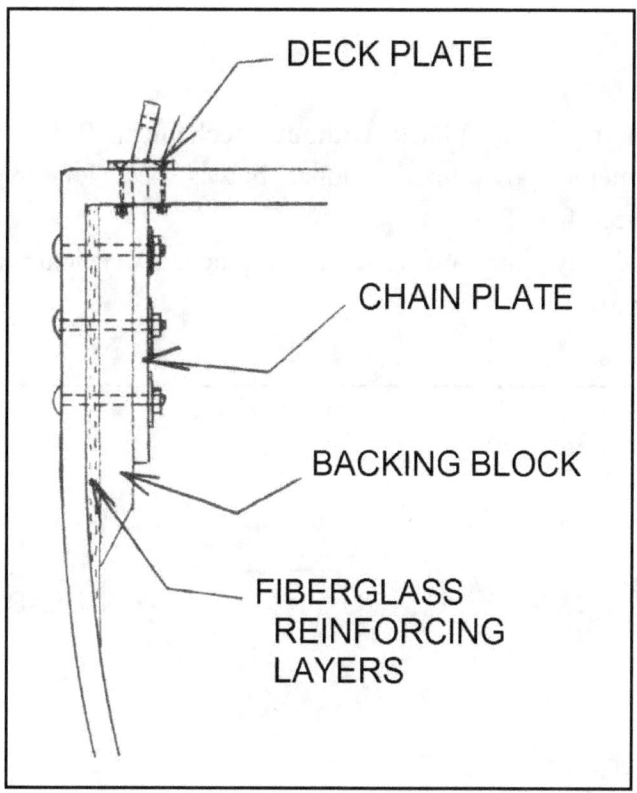

Internal mounting of chain plate.

should have large heads, and should fit snugly in the holes passing through the fiberglass. Wood pieces are fitted between the inside of the hull shell and the chain plates. Slots are cut through the deck or cabin top for the chain plates to pass through. Chain plate covers are commonly used on top of the deck or cabin top. These pass over the top of the chain plates and are bedded down and through-bolted to the fiberglass.

Replacing bulkhead-mounted chain plates is often relatively easy. A typical job is to remove old chain plates, cut larger slots through the fiberglass shell for the new ones, and mount the new chain plates in place. In some cases, old holes have to be filled in and new ones drilled. Often, the new chain plates extend farther down on the bulkhead.

Often, standard chain plates are used for backstays. Twin backstays can be added by installing chain plates to the transom in appropriate locations.

Bonding chain plates without mechanical fasteners is a secure method if properly done, but is best left to highly experienced fiberglassers.

Head-stay plates are sometimes replaced. A typical mounting is shown.

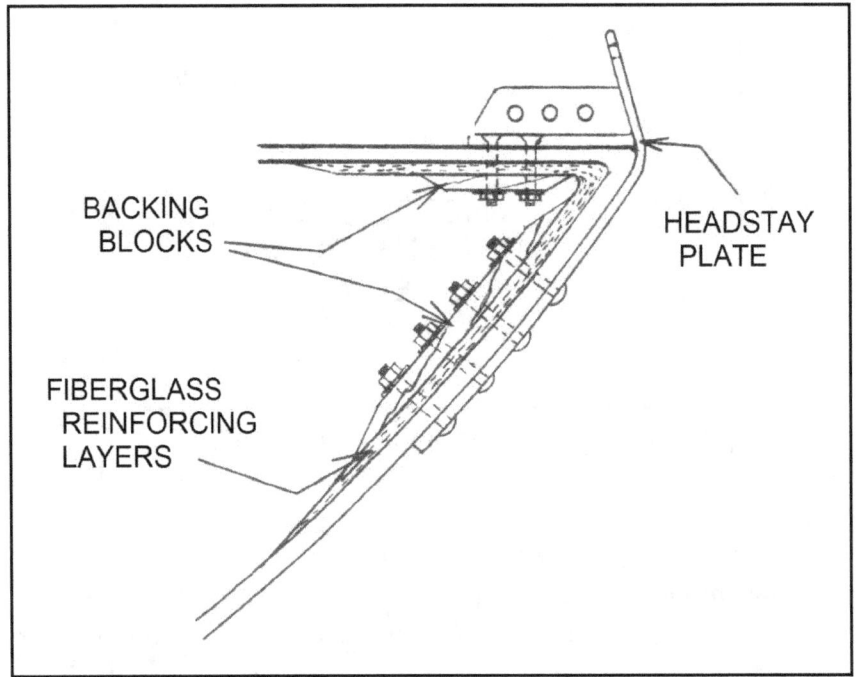

Mounting of head-stay plate.

Masts

Increasing the mast section is sometimes done, although this is generally an extreme modification. Check first with the manufacturer of the boat or with a designer, for this job involves adding extra weight in an undesirable location. The mast is best purchased with all major fittings attached. A number of firms now do this type of work.

Standing Rigging

The most commonly used wire is 1-by-19 stainless steel. Swaged-end fittings are typically used. High-pressure swaging machines are used to install these fittings. If you cannot have this done locally, a number of companies will make wire rigging to your specifications through mail order.

Another method of attachment is to use a thimble and compression sleeve. A hand tool is available for crimping the smaller sizes. A hydraulic press is used for installing larger sizes.

If the present rigging is to be replaced with a wire of larger diameter, the old rigging can be used as a pattern for making the new set. If larger turnbuckles are to be used, the size difference should be taken into account.

The turnbuckles can be of bronze or stainless steel. The stainless steel ones are generally preferred, but they are also more costly.

Once the wires and fittings are made, the remainder of the installation is generally easy. However, most amateur builders are limited in the amount of work they can do by the special tools that are required. Thus, it is generally necessary to have everything made at a company that specializes in this type of work.

Running Rigging

The running rigging consists of sheets, halyards, tackles, lifts, and so on. Many modifications in running rigging are fairly easy to make.

In many cases, the boat owner wants to attach a fitting such as a block, slide, or fairlead to the deck or cabin top. The basic method is through-bolting with a backing block. In some cases, additional reinforcement of the fiberglass shell is required. Bedding compound should be used on through-hull fasteners.

In some cases, you may want to change to another type or size of winch, or to add additional ones. A typical mounting is shown.

Wire halyards are typically made of 7-by-19 stainless steel, which has good flexibility for passing over sheaves. End fittings are generally installed with a high-pressure machine. For rope halyards, Dacron is generally used.

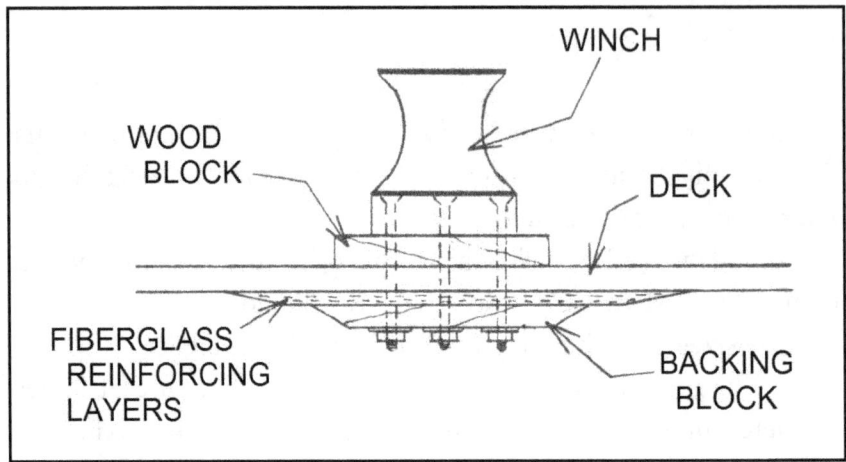

A typical winch mounting.

Gallows Frames

Only a few stock fiberglass boats come with gallows frames. Many cruising boatmen wish to install these on stock fiberglass boats. They can be installed in cockpits or on cabin tops. A typical installation is shown.

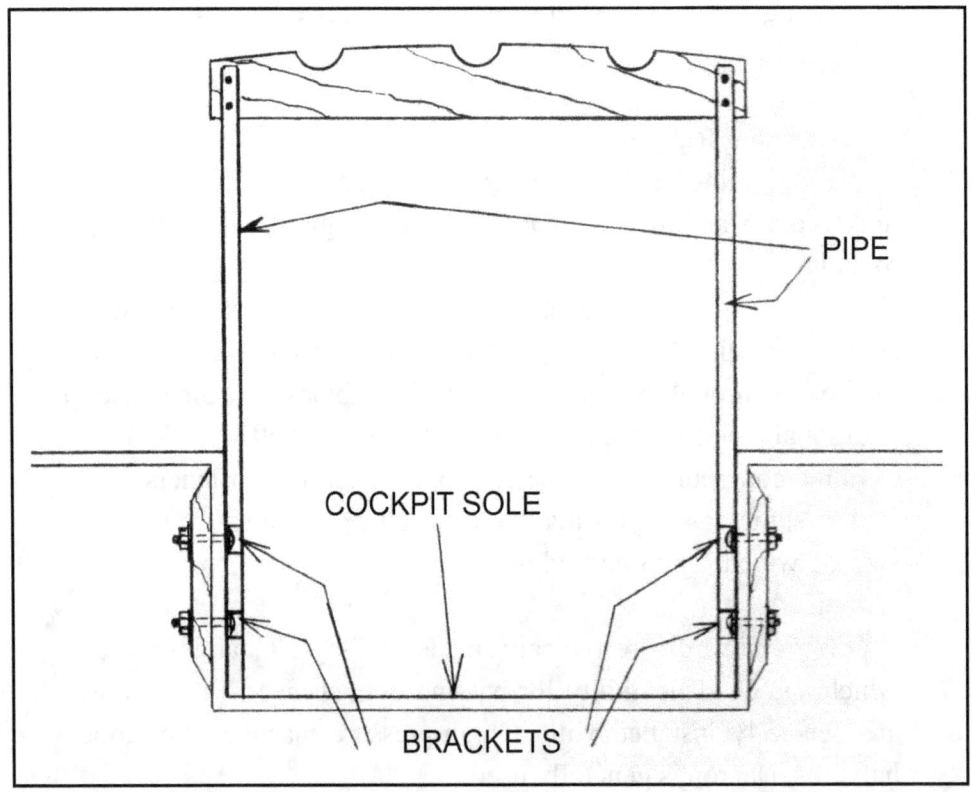

Gallows frame.

Roller Reefing

The roller-reefing units that are available fit most standard masts and booms. These generally come in kits with all necessary hardware. Installation is generally straight-forward. Modifications of the main-sheet leads may be required.

Jib Furling

A number of units are now on the market. In many cases, the present jib can be adapted. The standard system uses a wire. Rod systems are also available, but quite expensive. A plastic unit that is installed over the head stay is now being manufactured.

ABOUT THE AUTHOR

JACK WILEY is the author of fifty published books on a variety of subjects. Books that may be of special interest to readers of this book include *Boatbuilding from Fiberglass Hulls and Kits*; *Living Afloat: My Ten Years of Living aboard Small Boats*; and *How to Live Aboard Your Own Boat*. These books are available from Amazon.com in both printed and Kindle e-book formats. For more information about the author and his books, go to: **http://www.amazon.com/author/jackwileypublications.**